GULLIBLE TRAVELS

GULLIBLE TRAVELS

by

PATRICK CAMPBELL

ANTHONY BLOND

For Fred Tupper

First published in Great Britain 1969 by
Anthony Blond Ltd., 56 Doughty Street, London W.C.1
© Copyright 1969 Patrick Campbell
Made and printed in Great Britain by
W. & G. Baird Ltd., Belfast
218 51122 1

The pieces on the following pages have
appeared in the Sunday Times, and are
reproduced by kind permission of the
Editor. Page 24, 27, 30, 36, 39, 45, 66,
69, 72, 75, 78, 81, 84, 87, 90, 93, 96, 99,
102, 115, 118, 121, 124, 127, 140, 143, 146,
149, 152, 158, 161, 167, 176.

GULLIBLE TRAVELS

IMPURELY PERSONAL

COOLLY, THROUGH A GLASS EYE

DOLDRUMS IN THE JET SET

HOME THOUGHTS FROM A BROAD

Gullible Travels

'HELLO. Fred here. D'you want to go to Hong Kong?'

I was surprised by the speed of my double reaction. I knew, of course, I would accept instantaneously, but in the same flash of time I know I hadn't got a thing to wear.

I realised something else as well—that when women said they hadn't got a thing to wear they really meant it. I saw, with absolute clarity, that it wasn't vanity or the desire to show off or even social panic. It was simply that the nature of the invitation whipped them out of the world they lived in, or could rise to at a pinch, cancelling their entire wardrobe at one stroke, leaving them with literally nothing to wear, in the sense that everything they already had would prove to be insufferably hot, cold, light, heavy, bright, dark, thick or thin, promoting physical discomfort so intense that it would be almost a joy to stay at home.

I must have hesitated for a moment because Fred said, 'You trying to say yes?'

'Yes.'

'Fine. Thursday. We'll be in touch.'

He rang off.

Hong Kong. The steaming Orient. Punkahs? Small, smooth, almost weightless Chinese flickering like swift fish around the legs—through the legs—of huge, sweating, meaty, red-faced people from the West, bulging with mashed potato and stew. . .

I rang him back at the Airline Office.

'I'm afraid he's just gone out to lunch, sir. Could I help you?'

7

'Well, I was just wondering. Perhaps you know I'm going on this Hong Kong trip? I was just wondering if it was going to be very hot there?'

'I'm a fraid I couldn't say, sir. Will I have him call you back?'

'No no. It doesn't matter. Thank you very much.'

'Thank you, sir. Goodbye.'

I almost rang back again to ask her not to tell him I'd phoned to ask how hot it was in Hong Kong. In his job he travels the world—Tokyo, Mexico, Rio, Peru. He wears light-weight American suits. He's ready for anything, always. I didn't want him to think that I wasn't in the same cool condition. He probably had a lot of other things to do, anyway, arranging this trip.

Hong Kong. The mysterious East. The inscrutable Orient. Susie Wong? Little Lotus Flower of the First Water. Or was that Japan?

I had one fairly good dark suit, not too heavy, that would do for the evenings—and at that moment I realised I didn't know how long we were going to be there for. Three of my six white shirts were in the laundry, and wouldn't be back until Saturday morning, by which time I'd be on the other side of the world. If that was where Hong Kong was. Where was it—really?

I found a tiny world atlas at the back of a Pears Encyclo-paedia. On the flat it didn't look all that far to Hong Kong. Across Germany, Turkey, Iran, India, Thailand and we were there. Then I remembered driving from London to the Riviera and feeling that even London to Dover was nearly enough, and that afternoon the interminable straight roads of Northern France and by the evening I'd got as far as a place called Troyes. Over dinner I looked at the map. I'd done about a third of the journey, having been on the move since early morning.

Sitting there at the table I suddenly felt the size of an ant, but a microscopic ant, an almost invisible speck, occupying an incalculably tiny area of the world. For the first time in my life I truly realised that the world was round, that you could go on and on and on for days and weeks and months, passing limitless millions of people on the way, and finish up where you'd started from. Driving from Dublin to Galway or even London to Inverness gave you no feeling of the roundness of the world. You had to be on a Continent to begin to appreciate it.

Calais to Troyes looked to be about 250 miles. But Turkey alone was four times as wide and India twice as much again. It was impossible to think about it. All I could do would be to present myself at the airport like a parcel, and let the airline handle me from there. But wrapped in what? Three white shirts, a fairly lightweight suit, two pairs of flannel trousers and a hacking jacket. I was going to arrive in Hong Kong looking as if I'd walked there, with some tattered, sweat-soaked rags in a broken suitcase by way of a change.

I knew what sitting in an aeroplane, even from Dublin to London, could do to your clothes. Cigarette ash everywhere, coffee and drink spilt, probably fruit salad in the lap, everything crumpled into premature old age. How much worse, then, after passing over Germany, Turkey, India, Thailand—and another thought—getting out at oven-baked, tropical airports, swept by sandstorms, alive with mosquitos. Sweating, bitten, lumpy, shirt collar too tight, trousers with two huge and sodden bags on either knee.

Whatever else happened I had to keep one shirt unused, and the dark suit, in case there was some great formal dinner party when we arrived in Hong Kong. Perhaps I could wash one shirt in the lavatory on the plane. But they were poplin. They'd never dry and even so they'd have to be ironed. The Chinese were great laundrymen. I could hand the shirts to room service when I arrived, saying I'd like them back in half an hour. How to say it, to avoid all possibility of misunderstanding? 'Velly clean. Chop chop.' Ludicrous. And utterly ludicrous to reduce this great adventure to a lot of female fussing about clothes.

Madame thought so, too, when I told her about it that evening. 'Just go out,' she said, 'and buy some.'

Agitated as I was I succeeded in being patient. 'You know perfectly well I can't buy anything off the peg. I'm the wrong size.'

'Not all that extraordinary.'

'But I am!'

All my life I've had this problem, people thinking I can just go into a shop and buy a suit or a shirt or a pair of shoes. I know they're wrong, but they never believe me.

My trouble is that I'm a quarter of an inch under 6 feet 5. This might be all right if it was evenly distributed, but in fact it's nearly all leg. My legs are two inches longer than any ready-made trousers I've ever been in, so that the only way I can cover my socks is by lowering the seat between my knees. Furthermore, one arm seems to be longer than the other, because I've got a 2-inch drop on my right shoulder, and even the shorter arm is an inch longer than all ready-made shirt sleeves, and if I can't show a little shirt-cuff I simply cannot bear to be out in the thing.

At the other end I've got difficult feet. They're almost round, like an elephant's. Lengthways they're size ten and sideways size twelve. No manufacturer, in my experience, has ever brought off the feat of combining these two sizes of shoe into one. It is, therefore, agonising pinching or slopping along in huge boats that soon begin to turn up at the ends like Turkish slippers.

'You got a very smart pair of shoes about a year ago,' she said. 'That place in Brompton road. Try them again.'

'They took about six months to break in. I'm certainly not going to go hobbling around Hong Kong in boiling heat with my feet being squeezed up my legs into my knees.'

'If the worst comes to the worst you can always wear a loin-cloth.'

My anxieties redoubled themselves. 'We're probably going

to have to attend all kinds of cocktail parties and things on the Peak. That's where all the English people live, in shantung suits and silk shirts. It's like Sunningdale, only worse—'

'Oh, for God's sake,' she said. 'You're getting a lovely free trip to Hong Kong.'

'It'd be all right for you if it was you. You've got all kinds of cool dresses and things, but for an unfortunate man of my size it's extremely difficult—'

But we'd had this discussion too often before. 'What agony to be you,' she said, bringing this one to an end.

Later that night I came to the conclusion that in the circumstances this was very nearly true. To be thrown into a tizzy—the word struck just the right note of feminine hysteria—by having nothing to wear for flying to Hong Kong. It came, of course, from having lived too long on the thin northern edge of the world. Too many years in Ireland, a small dot in what often felt like the middle of the vast, grey Atlantic. An island as remote in every way from the bazaars of Calcutta, from the plains of Persia, even from the Riviera, as somewhere like Tristan da Cunha, equally drenched by drifting rain.

I was an islander, in thick trousers and a jersey and a mackintosh, looking upon driving to Dingle as an adventure, still waiting to be sure that the world was round.

By now I'd learnt that we were going to spend the week-end in Hong Kong. We would leave London on Thurday morning, arrive in Hong Kong the following afternoon and start back again on Monday—just a weekend on the other side of the world.

By the time I'd finished my shopping quite a number of gentlemen's haberdashers in Regent Street had been told about this bold adventure, but only one of them got the benefit of my actual custom. From him I bought a white, alpaca jacket, the sleeves of which surprisingly enough almost covered my hands. I also bought two very thin, floral sports shirts with short sleeves and a pair of black shoes made of woven, open-work leather thongs. None of them were anything like what I'd been hoping to find and, indeed, were so tasteless and gener-ally unpleasant that I knew if I were ever driven to wearing them in London it could only be with the accompanying explanation that I'd bought them almost for nothing in Hong Kong.

We took off from London and landed almost immediately afterwards in Frankfurt, where some more journalists came aboard. They scarcely had time to complain about having to pay for their own drinks before we landed in Vienna, to be joined by half-a-dozen Austrian newspapermen. This was an inaugural flight and the airline was intent upon getting the maximum publicity for it.

We left Vienna and landed in Istanbul, where for the first time I got out of the plane, wearing a white shirt and grey flannel trousers. It was like stepping into an oven. The air-field was a sandy waste, with a wind from a furnace blowing across it. Turkish soldiers wandered about, carrying machine-guns. The adventure was really beginning.

In the airport building—a large, battered shed, shimmering in the heat—I cashed a traveller's cheque for £10, receiving in exchange Turkish lira. The notes were thick and pulpy, almost like soft cardboard. I cashed the cheque because I'd already used up my English money, and wanted to buy my first drink in Turkey.

I had a cognac—seeing that the word was universal. It tasted of pepper and brown sugar. The barman gave me some smaller but even thicker notes in change.

Shortly after leaving Istanbul I asked the stewardess for a bottle of beer, and 'What's that in Turkish lira?'

'I'm sorry, sir,' she said, 'but we don't take Turkish money.'

I couldn't believe it. An international airline would surely accept any currency in the world.

'I'm sorry,' she said, terminating the discussion at one stroke, 'but not Turkish.'

In the bar in Istanbul I'd felt myself easing pretty comfortably into the role of world traveller, but suddenly now I was right back where I'd started—a panic-stricken islander, penniless and a thousand miles from home.

I examined my Turkish money again. The printing was blurred and messy, even the colours had run. It looked as if the barman could have printed it himself. I wondered if I could change it into Lebanese money in Beirut, the next stop. But then I'd have about £9 worth of Lebanese whatever-they-were which wouldn't be acceptable in Karachi.

All at once, and quite clearly, I saw myself going mad, cashing traveller's cheques at every stop and finishing up in Hong Kong with a dozen different kinds of currency that not even the hotel would accept.

Social anxieties about having nothing to wear had given way to the straight terror of not being able to get anything to eat.

I sat huddled in my window seat, looking down at the brown whorls of the deserts of Arabia, knowing that I was really only up to travelling to Dingle, and even at that I'd want to be careful to keep my head when I got there.

I was saved by the man sitting beside me. 'If you're really lumbered with all that Turkish stuff,' he said, 'I can lend you a tenner till we get there.' In excessive gratitude I tried to give him a traveller's cheque, but he waved it away.

'Too complicated,' he said. 'Just give me the cash in Hong Kong.' He knew his way around the world. I'd seen that his wallet was full of pounds and dollars.

We landed in Karachi, and into poverty the like of which I'd only seen before in the West of Ireland. Paint was flaking off the bus that took us to the airport building. It had lost a wing and one of the tyres was nearly flat. Litter was blowing around everywhere. One of the windows in the bar was broken. The barman's white coat looked as if he'd been delivering coal.

I ordered a cognac for myself and for the man who'd lent me the money. The barman poured it from a Hennessy bottle. It tasted of pepper, Tabasco sauce and brown sugar. 'That,' he said, 'is one and one half English poun.'

'Well,' I said to the other newpaperman, 'I certainly shan't be coming here again in a hurry.'

He scarcely bothered to smile. He probably knew that a teaspoonful of alleged brandy cost 15/- in the airport bar at Karachi, and would have had a bottle of beer if left to his own devices.

Through the night we flew across India to Calcutta. I slept for a few minutes at a time and then woke again to listen to the roar of the plane. After Hong Kong it would go on without us to Tokyo, Honolulu, San Francisco, New York and back to London, where it would begin again. Frankfurt, Vienna, Istanbul, Beirut . . .

It was no longer possible to doubt that the world was round. Also, there was evidently a very great deal of it.

I stayed in the plane at Calcutta. It was too early in the morning for cognac and to drink tea in India seemed to be altogether too much of a good thing.

Later that morning we landed at Bangkok. The fence outside the airport building was lined with hundreds of tiny Thailanders, none of them seemingly more than five feet tall. They giggled delightedly as they saw the scruffy Western giants tottering down the steps of the plane. In the airport building the minute Thailand air hostesses with their little golden faces and lilac coloured uniforms were almost edible, two or three to a mouthful, beautiful beyond belief.

I sat down carefully in an armchair in the lounge, not daring to move about, feeling myself to be—in comparison with these small, quick, golden little people—hopelessly uncoordinated and of a truly monstrous size. The heat was stupendous. Sweat dripped even from the ends of my trouser legs, while the hurrying Thailanders remained as crisp and as fresh as ginger biscuits.

It was even hotter in Saigon, when we landed there a couple of hours later. This was before the Americans had entered the war and the whole country had been enveloped by it, but nevertheless there was a feeling of jumpiness around. A number of people ducked when a car backfired, and none quicker than myself. Once again I was distressed by my huge Western size. Standing so high above the Vietnamese I felt like a target that no one could miss. Many Americans were later to feel exactly the same way.

We flew on over the China Sea, and began to lose height. After thirty hours of flying we were nearly there. All at once the sea was covered with mountainous but heavily wooded islands. As we came down still lower the trees were all round us and even above us. The plane rocked about, then steadied itself. Just ahead was the long runway, a single strip of concrete running out into the sea. Everyone held on to something fairly tightly as the plane put down. I gathered afterwards that it's always a pleasure to land without incident in Hong Kong.

Outside the sirport building there was a bus with our destination written along the side of it—The Peninsula Hotel. The Chinese driver took us at a fearful speed through streets alive with millions of people straight to the Miramar. After a splendid altercation with nearly everyone in the bus he drove us very slowly to the Peninsula.

It was plain that he still didn't believe it.

The Chinese commissionaire said, 'How are you?' So did the receptionist. The liftboy said, 'How are you?' too. And so did the pageboy who carried my bag. The very old Chinaman whom I found in my bathroom said nothing. He seemed to have some official purpose there, because he had a bucket and a mop, though he wasn't using either of them.

We looked at one another in fear for quite a long time. He said something in Chinese, pointing to the bath with his mop. I thought he might be some kind to bath attendant, waiting to get to work on me. I shook my head and pointed to the door.

He pointed to the bath again, with his mop. I shook my head and on a sudden inspiration gave him one of the smaller notes from my Turkish store. He looked at it, gave it back and left the bathroom, muttering.

I turned on the hot tap. Not a drop of water emerged. The cold one was also dry. There was no doubt that I was face to face with the mysterious East, and with no means of finding out what it wanted, or was prepared to give me.

Next morning I put on the white alpaca jacket and the open-work shoes. I looked like the driver of a coach on an outing to Folkstone on August Bank Holiday. I put the jacket away, never to wear it again, and put on one of the new floral sports shirts and sauntered out on to the Nathan Road to do some bargaining.

Alternating with the shops selling jade and jumpers decorated with diamanté were others dealing exclusively in money. Their frontages consisted of a small brass grille, with a face behind it, surrounded by large advertisements for rates of exchange. I pushed my Turkish lira through the first grille I came to. It was pushed straight back again. A Chinese face said, 'Cullency too soft,' and beamed as though it were the best news it had been able to hand out all week. I tried it again, next door, with the same result, and decided to abandon further horse-dealing on money market.

At midday I had an appointment to meet a distinguished Englishman at the Hong Kong Club. I crossed on the ferry with a million Chinese and on the other side pushed my way through two million more. Already, in Kowloon I'd seen half-finished blocks of flats occupied by swarming Chinese families, with thousands more waiting in the street for the next floor to be put on so that they, too, could move into the empty shells.

I was refused entry at the door of the Hong Kong Club. The hall-porter said, looking at my floral sports shirt, 'No tie. Go please to bar at back. How are you?'

When the distinguished Englishman arrived he was wearing

a beautiful white shirt, a regimental tie, a beige silk suit, boned brown shoes and a Panama hat. The temperature was 102 degrees.

After a jolly week-end I left the mysterious East on Monday morning and after the now familiar kangaroo leaps around the world arrived at London Airport the following afternoon, wearing the cleaner of my two florals.

There was a bitter wind blowing, accompanied by gusts of freezing rain.

The taxi-driver looked at me. 'Where you bin, then, mate?' he wanted to know. 'The good old Costa Brav?'

'No,' I said. 'Just a weekend in Hong Kong.'

He closed an eye. 'And I'm Susie Wong.'

He might well have been, for all I'd seen of her.

Next day I took my Turkish lira into the bank, £10 worth of them, less the price of the cognac in Istanbul.

After interminable deliberations, much of them conducted out of sight somewhere round the back, the cashier gave me £3 15s. in English money.

Damn!

ONE is not, of course, a libertine or anything excessive like that, but the sight of this delicately matured piece of crackling, sitting all by itself in the corner, was so arresting that I gave my hostess but the most perfunctory of greetings before weaving swiftly through the crowded room to slot myself right in beside it.

'Good evening.' I said. 'And who are you?'

Not a second wasted. Getting the spade into the work without a second's delay.

The lady turned to look at me through lowered lashes—admittedly false but lushly cemented into place. Perfect heart-shaped face, framed by smooth blonde hair, shoulder length, something after the style of Rita Hayworth. But, brother, ever so more so.

When she spoke it was in a voice so husky that I came within an ace of bursting into applause. The content, however, was a little lowering, even a little abrupt in tone.

'*Parle pas Anglais,*' she said.

Something of a setback. After all this time in France a fair measure of badinage is available to me in the French language but the more delicate subtleties, the telling and elegantly turned subjunctive, the suggestive but gracious compliment rather tend to elude me. The level rather closer to, 'What about it, duck?' than, 'I drown myself in the twin pools of your mysterious, moon-stone eyes.'

However, the lady had said, rather curtly, that she didn't

18

speak English, so that it was up to me to flash a subjunctive or
two in her own language.

I asked her if she was alone. '*Mais non*,' and she indicated
with a soft white hand a distinguished looking elderly gentleman
on the other side of the room. She was wearing a wedding ring.

'Your husband?'

'Not exactly,' she said, in an absolutely level voice.

In fact, the few remarks she had delivered had all been in
this curiously unemotional form. Nor, after the first moment,
had she looked me in the eye. A less sanguine disposition might
almost have come to the unthinkable conclusion that the lady
was bored, and so bored, indeed, that if I didn't leave in the
next couple of seconds she would.

I decided to pump some steam into the atmosphere. 'I do
hope I don't disarrange you by talking to you like this without
an introduction,' I said, 'but the fact is that there are very few
beautiful women on the Cote d'Azur, particularly during the
winter months.'

The lady produced a cigarette. I lit it for her. Once again
she didn't look at me, accepting the service without noticing
it, as though I'd been a manicurist, working on her nails.

'The French girls mostly have very long noses and very short
legs,' I went on. 'In fact, their legs are so short that their knees
seem to be joined on to their bottoms. I can't imagine what happens when they try to run.'

This mildly risqué anatomical chat continued, to my regret,
to leave the lady as unmoved as ever. You could even say that
things were getting worse. Through the smooth blonde hair I
could see the lovely line of her jaw tighten for quite a moment
or two, as she held back what could only be a yawn.

The only thing to do was to put my foot right down, drive the
accelerator through the floorboards. So pointed a brush-off
was not to be borne, after so little provocation.

I leant in a little closer. 'As, at the moment,' I said, 'you're
sitting down it isn't possible to check the facts with a hundred

per cent accuracy, but from what I can see of it from here your knees rather definitely aren't fastened to—your knees are not attached in this unfortunate fashion.'

At last the lady turned to me, but there was no warmth in it. Only disdain, like a duchess unforgiveably importuned by the bootboy.

'*Non*,' she snapped, disposing in one word of a filthy suggestion, and all further conjecture about it.

At that moment I found my wife was standing above us. I struggled to my feet and, at a loss, said, 'Darling, this is Madame—', and had to leave it at that.

Madame C. looked at the other one for a brief moment. '*Bon soir, Monsieur*' she said, and moved on.

Like a lot of coal falling downstairs everything dropped slowly into place.

The other one was a chap.

Hop, Skip and Jet

SOMBRELY, the mature playboy examines his reflection in the full-length mirror.

This is his night to howl, rave, freak-out and have an enjoyable time.

A soirée is in prospect with the International Hop, Skip, Jump and Jet Set. There will be a Contessa or two, some Virginian polo players, a swatch of top models, one of whom wears transparent trousers and nothing else, a bunch of *avant garde* French film makers, a convention of Argentinian drug peddlers, and, more than probably, Lionel Bart. Nothing, really, too much out of the ordinary, but cool enough all the same.

With increasing gloom I—for the mature playboy is none other—continue to study my image in the mirror. And what an image it is. Stodgeville, man. The distilled essence of Creepington.

The disappointing thing is that only five years ago this freaking-out night-time clobber of mine was pushing up towards the Number One position in the charts. Nothing as swinging had been seen even in Notting Hill Gate, where I had my residence at the time.

A single-breasted satin lapel and faint stripe—a faint *stripe*, if you please—in the material of the dinner-jacket itself, so that it seemed to shimmer as I passed through the motions of the Twist.

Actually, I had the devil of a battle to get my chap to run up

the ensemble at all, especially when I asked him to cut the
drawers a bit high in the front, as I proposed to wear a moiré
silk cummerbund, if I could find one that would meet around
the back. He knuckled under in the end, but got his revenge by
making the jacket so voluminous that even when it was buttoned
I could see my feet inside it by leaning very slightly forward.
Extraordinary chap, with a mind, apparently, of his own.

But now, faced as I am with the Hop, Skip, Jump and Jet
Set, this once way-out creation looks as though it would have
been rejected, as unsuitable for the occasion, by the Master of
Ceremonies at a small staff dance in Northallerton, in 1929.

Something has got to be done, and particularly around the
upper bust. Distastrously, you see, in 1968, I am wearing a
pleated shirt with a neat black bow-tie, when in 1968 it should
be blouses, blouses all the way. But where can I find a blouse,
or even a white silk roll-top jersey at this hour of the night?
My wife, admittedly, has ammunition of this kind in plenty,
but then—even around the upper bust—I'm twice as big as
she is.

But wait! Did I not read somewhere—quite recently, as
early perhaps as yesterday—that the really swinging young
Blades have abandoned the blouse and the jersey for evening
wear, and taken to the white silk choker! I haven't got a
white silk choker, but I have got a very large red handkerchief
with white spots that someone gave me, having run short of
all other invention, for Christmas.

I knot it round the gullet, and man, it's Winnersville! But
now the jacket is all wrong. Too conventionally single-breasted
to be borne. But suddenly the intransigeance of my tailor comes
to my rescue. The voluminous single-breasted, with a safety-pin
there, and there, and there—buttons to follow later—can be
transformed in an instant into the likeness of a full-dress
double-breasted uniform jacket, just the thing for relaxing in
after the Charge of the Light Brigade.

I'm getting excited. If this keeps up even Li Bart will be asking me where I get my schmutter.

We are, however, not home yet. The trousers, unfortunately, still match the jacket, and I'm certainly not going to go raving with the Hop, Skip, Jump and Jet Set carrying that sort of social burden. That's Moansville, man.

Trouser-wise, are not giant plaids, in velvet, the sine qua non for evening wear? I am deficient in giant velvet plaids but on the other hand I've still got a pair of grey, Glenurquhart check trousers that will certainly last out the evening, provided I don't lose my cool.

A smash! Now, rapidly comb the hair down over the ears, ignoring the rather empty, domed portion on top, and I'm ready for the ball.

A final check in the mirror. A small voice warns me I look not unlike a tramp who's been unlucky with his hand-out from Lady Bountiful, but I ignore it.

I have the comfort of knowing I'm going to look just like everyone else.

I'm Certainly Glad to Meet You

I KNOW I'm awake because I can feel I'm lying flat on my back and looking up at the ceiling.

I know I'm awake because if I was asleep I'd be dreaming and, on the record of the last few weeks, if I was dreaming—lying flat on my back— I'd be lying on broken molten glass and if I was looking up at the ceiling it would be just in time to see it suddenly covered with human gore, and screaming faces about to fall upon me and tear me to pieces with their poisonous, pointed fangs.

But the ceiling is smooth, grey and motionless and the mattress is soft beneath me so I must be awake. But only just, and I want to keep it like that for as long as I possibly can—or I can get things a little straighter in my mind.

This morning—and it must be still very early—I have a sense of foreboding. I have a weight lying upon my soul heavier and darker than all the lead in the world, all put together. Yet it's not true to say that this weight is lying upon me. It's rather more suspended just above me, waiting to fall upon me like a whole block of flats a split second after I'm able to identify its nature.

I keep my eyes narrowed in slits as fine as razor blades, keeping myself well below the level of full consciousness, but at the same time feeling around with a gossamer-like delicacy to try to remember what I was up to last night.

I've got a kind of feeling that last night was the culmination of months of frantic, almost hysterical endeavour. I've been running myself clean into the ground in pursuit of some honour or promotion which other people want for me much more than I want it for myself. I've got a memory of endless meetings and myself talking and talking, arguing and persuading, making more and more desperate efforts to convince. But the wall that faces me seems impenetrable, as solid as the Rock of Ages. . . .

Good God! It couldn't be true! But it's mad enough.

Dont' tell me I've been trying to shift old Avery Brundage out of the top job on the International Olympics Committee, and snitch it for myself! And got it, too!

For a moment it seems just possible. For months back I and everyone else have been yelling about Black Power and drugs and nationalism and police brutality, and if that isn't the Olympic Games what is?

I arch myself up on my elbows, aghast at my own idiocy. Only another four years to Munich and I still haven't decided whether the Olympic Flame should be carried in an armoured car. . . .

Yet it doesn't feel right. I don't give a damn about the Olympic Games and in any case by 1972 every nation will be holding their own, behind barbed wire on their own territory, leaving old Avery and the Committee ploughing away in limbo. No. It couldn't be for that that I've beat my brains out month after month after month. It was for something far worse.

I lie flat on my back again. The monstrous weight trembles. It is about to fall. I turn half on my side, to dodge the shattering impact, and my eye falls on my wife, in the next bed.

The weight shivers again. She's in this, too. Whatever I've been at she's in it with me, up to the neck.

She's been talking about new clothes. 'Do you think two hundred full-length ballgowns would be enough?' She's been on about moving. 'Are sheets and things provided, or do we have to bring our own?' She's also gone crazy about culture.

'What sort of culture do you think we ought to have? How many people play chamber music at the same time, and will they want tea or coffee at the interval?'

As if I hadn't got enough to worry about. As if I didn't—
I rise straight out of the bed, horizontally, screaming,

As if I hadn't got enough to worry about. As if I didn't— scrabbling at the air with arms and legs.

It isn't a matter of didn't. I have.

What I've done is to go and get myself elected President of the United States of America! That's what I've gone and done.

I fell back on the bed, from the ceiling, whimpering, dragging the bedclothes over my head. The monstrous weight has fallen. Everything has come clear. I'm done. I'm the next President of the U.S.A.

But suddenly my very soul is suffused with the gentlest, balmiest kind of well-being. I hear silvery music. I am wrapped in silks and velvet. I've just realised—

IT WAS THE OTHER FELLA WHO GOT THE JOB!

The Evil Furry-faced World

THE other day I stood briefly framed in the garden gate, about to step prudently out into the road, when the most awful looking man you've ever seen in your life blew his horn at me.

He didn't even meet my eye. He simply sat there behind the wheel of this large, sporting saloon, wearing a fearful moustache on a creamy complexion made of ivory soap and a spotted scarf round his neck and as he swept past a couple of inches away from me he blew his horn once, imperiously, 'BLAH!' and went on down the hill.

When one is in the middle fifties and worried about American isolationism and British decadence and racial unrest any sudden sound can cause the heart—(another worry?)—to speed its action to bursting point. When, however, a man with an awful black moustache, who you think is going to pass quietly by, suddenly lets off his horn straight into your face, it nearly always causes people in the middle fifties to make a small, involuntary spring into the air, and if they were standing on a step before they are likely to come down badly, slightly twisting an ankle and even grazing an elbow on the stone gatepost.

I achieved a fairly full bag of these minor injuries and was consequently only able to shout at him after he'd gone round the corner and was probably out of earshot.

I sat down in the potting shed to recover, and to make a short list of what was wrong with the world.

(1) The infinitely tedious and omnipresent Beatles, whose idiotic moustaches have furred mankind everywhere.

(2) The general tendency towards absenteeism, causing a man under thirty with a drooping black moustache to be abroad on a country road at 11 o'clock in the morning, showing no perceptible traces of guilt, while wearing a spotted scarf.

(3) The universal craze for speed, which drives idle young men with drooping black moustaches to plaster their cars with GT signs, plastic ribbons, streamlined wing mirrors and hooded spotlights, and to blow their fortified Monte Carlo Rally horns at persons of modest demeanour, standing innocently within their own gates.

Warming to my work, I was about to advance into Clause (4) —the Churches' failure to combat hedonism—when, unbelievably, the fortified horn went off again.

'BLAH!' it went, as it passed the gate, presumably in case I was sufficiently recovered to venture out again.

This time he had a young woman with him—a veritable Jezebel of about eighteen. They were laughing, almost hysterically, together—if not already drugged at least morbidly stimulated by the prospect of taking same. Just as they turned the next corner, still laughing, the spare wheel emerged mysteriously from the boot, bounced once and lay on its side, quivering a little, in the middle of the road.

Forgetting about the first three clauses, and the partially formed fourth, I was good enough to cry after them, 'OY!', but they'd gone.

It was a beautiful wheel, racing alloy, with a brand new wide-tread tyre. About thirty quid's worth, or double if he got punctured on a wet night without it.

I propped it inside the fence, near the gate, and thought.

(1) My garageman, with a keen interest in sports car, would be glad to have it, taking a corresponding gusset in my bill.

(2) The motorist might just have seen me, in his rear mirror,

picking it up, and be looking forward to collecting it on his return.

(3) The whole thing a generally uncertain proposition, leading if things went wrong to probably adverse legal action.

(4) Leave it alone.

Next morning the wheel had gone, together with our step-ladder and a lot of the builder's cement, wheelbarrows and tools.

For nearly a year he's been trying to build us a non-smoking fireplace, but yesterday he gave up, saying that next week he'd begin all over again, with an altogether different design. He left me with a number of problems.

(1) A telephone call revealed that he'd gone on ten days holiday.

(2) Did one of his workmen put the spare wheel and our step-ladder into the van, thinking they belonged to the firm?

(3) What will happen if the motorist comes back, looking for his wheel?

(4) What on earth is the good, in this materialistic, hedonistic, furry-faced and speed-crazed world, of trying to do a good turn to anyone?

Clarence House—At Last!

MY eye flashed down the third column on Page 3 of the Sunday Times last Sunday, rapidly misreading as usual the smaller headlines.

Thus 'AIR SHOW CRASH FILMS FOR INQUIRY' became 'AGE SHOW CRUSH FILMS FOR ENQUIRERS' and 'ROAD CLOSED BY FLOOD THREAT' turned into 'ROAD CLOSED BY FOOD THREAT'—perhaps a mass distribution of sausage rolls by the Granada service station on the M.4. which had put the great motorway out of action.

Then, immediately below this, I saw that I had succeeded the Lady Jean Rankin as Lady-in-Waiting to Queen Elizabeth the Queen Mother.

The hurried eye took in the brief paragraph which comprised the Court Circular, and there it was, clear for all to see: 'Mr. Patrick Campbell has succeeded the Lady Jean Rankin as Lady-in-Waiting to Queen Elizabeth the Queen Mother'.

A moment later, of course—and an anxious one it was, too —I saw that it was really someone called Mrs. Patrick Campbell-Preston who had succeeded the Lady Jean Rankin, but the shock lingered on for quite some time. It lingered so long, indeed, that in this state of alarm I began to make plans for taking up my new appointment, hampered in no small measure by my present exile in the South of France.

When this kind of major posting comes the first necessity is to try to find out why one has got it, in preference to anyone else. Only in this way can one enter upon one's new duties with the confidence required for their proper execution.

On the other hand, other important factors are also involved, like the strong possibility of becoming subject all over again to British tax, the repurchasing of a domicile within handy distance of Clarence House and, speaking of Clarence House, what the Lady-in-Waiting would have to do there when she (he) was in it.

The agenda seemed to be so full that I decided to take first things first: i.e., to find out why I had been chosen as the first male Lady-in-Waiting to the Queen Mother. Evidently, so far as the Royal Family were concerned, my qualifications must have been of the highest, seeing that there must have been any number of female ladies queueing up for the job.

The obvious thing to do was to give Philip a tinkle at Buck House, presuming upon having lunched with him on several occasions some years ago at an organisation called the Thursday Club. Presuming? Presuming nothing. If the new Lady-in-Waiting to a chap's mother-in-law can't ring him up and ask him what's the score, the whole social élite has gone to pot.

But how to go about it, without having the telephone number to hand?

'Je voudrais parler personnellement au Duc d'Edinboourg, chez le Palais de Boockingom, a Londres, mais malheureusement je connais pas le numéro.'

And then the imperious one at the local exchange barkink out, as usual, 'Comprends pas! Faut parler Francais!' As if I wasn't.

All much too difficult. Better just to accept the honour graciously, almost indeed as though I had been expecting it for some time and had even been wondering what had been keeping them at Clarence House.

Perhaps, the financial side? A discreet telephone call to my predecessor, the Lady Jean Rankin, to ask her what the screw, if any, was like, and the perks—also if any, including possibly free accomodation, with meals of course, in the attic.

Little point in that either. The post of Lady-in-Waiting to the Queen Mother has an honorary flavour about it, suggesting that you've got to bring your own loot. And very right and proper, too. But would my accountant concur in this opinion? After all his trouble in getting me out would he regard it as a fiscally viable proposition to find me once again on the back of his neck, admittedly in a highly exalted position but one producing the absolute minimum in the way of scratch?

'Hello? Mr. Ive? Look, something rather odd has just turned up. Perhaps you've read about it in the papers? No, no. Not in the nick here. As you say, not at the moment. But, look, I've just been appointed Lady-in-Waiting to the Queen Mother and I've been wondering if it would be a fiscally viable proposition for me to—Under no circumstances? Certainly not? Thanks a million. I was just wondering . . .'

What a relief not to have to wonder any more, to come out of the state of shock, and to wish Mrs. Patrick Campbell-Preston the best of British luck in her new job.

Nostalgia Here I'm Not

I was having a brisk turn-out of my filing cabinet the other day, looking for some bank statements which the bank stated that I had, though of course I knew quite well that they had them or that, alternatively, I'd torn them up two years ago, when—under a folder labelled 'TOP PRIORITY MISCEL-LANEOUS GARBAGE'—I came upon a photograph.

It was a studio portrait, turning a little yellow, one of those studio portraits from an earlier age, when Dorothy Wilding was taking her snaps with the light behind the subject's auriole of hair. The whole face, indeed, was lit by a kind of ecclesiastical radiance, its delicate contours even further softened by the photographer's art.

I looked upon it for some time with an emotion approaching love. The whole of youth was there. The shy and tender lips, the innocent, appealing eye, the general appearance of gorm-less idiocy. I sighed, sweetly and gently, for the photograph was of none other than my good self, taken some thirty-five years ago.

A deeply moving experience. Yet, beneath my justifiable emotion, something else was demanding my attention, calling upon me to identify another element in the picture, above and beyond or, perhaps, slightly to one side of, the bashful inno-cence of my own face.

I got it almost at once. Taking it by and large, I looked exactly like Twiggy! Not feature-wise, of course, but the impact of the face was the same. The same softly waving hair, the

33

B

same flat, impassive eye. The cricket shirt I was wearing even looked like one of the clinging new rayon blouses. There was I, indeed, the very spirit of the Thirties itself! Except, of course, that I was real, and that it had probably cost Twiggy a couple of hundred quid in hairdressing fees, make-up and some hard work on the part of her couturier to give her the same look.

The spirit of the Thirties. I tried to remember what it had been like. After a lot of aimless and abortive dredging I came up with one fairly clear recollection. Round about 1932 I owned a light mauve shirt, with two detachable stiff collars in the same colour. The collars were very high, and the purplish tie I wore with them had a very small knot, giving me—I felt sure in retrospect—the appearance of a victim of unexpected strangulation.

Memory began to work more surely, more swiftly.

I once bought two tickets (5/-) for a dance at a tennis club outside Dublin, intending to escort a girl called Dymphna to the rout.

I began dressing at four o'clock in the afternoon, giving myself plenty of time. I was ready by 5.30. A preparation called Anzora Viola (Masters the Hair) had set my waves like concrete. I wore my mauve shirt and the high and stiff mauve collar and the purple tie with the tiny knot. I wore a pair of almost white flannel trousers, which I'd pressed myself, and black patent leather dancing pumps. In place of a more formal jacket I put on an Old Rossallian blazer, in stripes of red and white and blue. In the breast pocket of the blazer I perched a clean white handkerchief and, after a final confident look in the mirror, went downstairs to my Morris Cowley—(£12 10s., plus a nearly new bicycle traded in.)

I met my father in the hall. He was coming in from the garden, fairly well covered in mud. The contrast in our appearance seemed to strike him with some force. 'Godstrewth!' he exclaimed, and stepped on one side, wide-eyed, to let me out.

Dymphna, when she opened her door to me, seemed equally

surprised, but not perhaps as favourably as my father. In fact, she said, 'What are you dressed as, at all?'

'It's a tennis dance,' I said. 'Semi-formal.'

'You look as mad as a hatter,' she said.

I thought she looked quite like a sofa, in red taffeta, but didn't say so. On the way there we had two punctures, one after another, in a matter of seconds, so that the car became useless. After some irrational complaints Dymphna got out and took a tram home. I went on to the dance myself and got thrown into some bushes by a couple of older men.

That, dollies, was the spirit of the Thirties. Rough, tough, hard sledding all the way.

Be thankful that you can achieve the Thirties look, without having had to endure them.

A Car was Cleaned To-day

THERE was an aura of instability about the lady. I could see only the back of her head through the rear window of her car, but even her hair looked uneasy, as though it had been subjected to some experimental cutting that hadn't quite worked.

Also, she was busy, in a quietly desperate way, trying to fix one of the huge new residents' parking confusions to the windscreen.

As the next customer disappeared into the maelstrom of the car-wash emporium I gave her a little toot on the horn, inviting her to move up in the queue.

It had a bad effect. The parking thing fell off the windscreen, the lady shot a terrified glance at her driving mirror, found it was out of alignment, threw up a hand to adjust it and dropped what was probably her handbag on the floor. At any rate, she disappeared downwards, from sight.

The man behind me gave an impatient blast on his horn, so I gave the lady another little tootle on mine. The difficult looking hair suddenly appeared again, jerking from one side to the other as the lady looked for the starter, which had evidently moved since she'd last seen it.

The man behind me gave me two blasts, I gave him a couple of digits in reverse, the lady's car, in a series of short, compact leaps entered the maw of the car-wash place and halted just in time in front of the two enormous, woolly rollers.

As usual, the laundry staff had altered completely since the previous week. In place of the three cheerful Africans there were now a couple of indolent white natives, round about seventeen years of age. One of them, in rubber boots, pressed the button and the rollers began to whirl, flinging out sheets of water. The lady sat there, in her car, without moving, terrified of the spray.

The lad in the boots switched off the machinery. 'Come on in, missus,' he bawled, waving his arms. He switched on again. The lady, blinded by the water, drove straight into one of the rollers, instead of between them, stalled her engine and came to a halt. She became invisible in the heart of the cascade, with her radiator jammed up against the whirling roller.

The booted boy switched off again. In a frenzy of indignation he pounded the windscreen of the lady's car, once again waving his arms. She opened her window, and they held a short conversation. At the end of it the boy pushed the car back, away from the roller, lined it up straight through the window, switched on again and then let out a great cry of, 'MISSUS!' The lady, edging forward into the spray, almost had time to shut her window but not quite. It must have caught her in the ear, because I saw her shoot up a protective hand before she disappeared into the Niagara.

When the boy switched off again I saw that the lady had got through to the other side, and was trying to make adjustments to her sodden hair, in the driving mirror. The lad, on the other hand, was leaning against the wall, holding his stomach in an agony of laughter. After a while, he recovered, and both lads soaped down the lady's car. While I passed through the the rollers she survived another deluge of water, this time apparently without damage, and moved on into the drying tunnel.

I had a clear view of what happened next. The lady, probably seeking instruction, opened her window again and got, straight into her affected ear, the full blast of the hurricane generated by the machine. Among the other things in the car that rose into the air and whirled around like demented birds was her residents' parking placard, which fastened itself grimly to the back of her neck.

Both lads lent against the wall, this time, sobbing with laughter, their knees visibly giving way.

The lady got her window closed. She sat at the wheel, her head bowed over it, her very soul soaked in water and shredded by the recent typhoon. After nearly a minute she recovered enough to put the car into gear. It leaped forward out of the tunnel, took another buck-jump into the forecourt of the garage and stopped stone-dead, the engine once again stalled. The lady made gestures indicating that she was unable to re-start it. The two lads, holding on to one another, staggered forward and gave her a push. The engine started with a roar. The booted lad, already weakened, fell down. The lady, perhaps with the intention of thanking them, threw open her door. It struck the fallen lad on the head.

The lady drove away, disorientated beyond recall. That evening, her husband probably said brightly, 'I see you had the car cleaned today.'

In the Home of the Gnome

A SMALL, discreetly opulent restaurant, with the lighting striking exactly the right note between intimacy and being able to see what you were at.

It had been planned for its precise purpose. The tables had been placed, scientifically, so that you could speak above a whisper and still not be overheard by your neighbours, while at the same time, without straining your voice, you could let them have an earful of what you were talking about, if such a strategem happened to be part of the master plan.

Even the maître d'hotel was perfectly cast for his job. Old, rolypoly, red-faced, slightly tetchy in manner—a man who'd heard it all before, who'd seen them come and seen them go and whose only remaining interest in his customers was that they should enjoy his excellent food and drink. It was plain that one could speak freely, while he was bending over the table, without—as it were—rocking the boat.

The serving itself was done by a young girl as carefully chosen as the rest of the ingredients. Fresh-faced, innocent, supremely competent, her sole desire could only be to do her job as well as possible, with a view to laying aside a nest-egg for her marriage to a reliable young man. No eavesdropper, she.

'Thooper thet-up,' I said to Madame, as we sat down.

She silenced me with a financial look—for where should we be but right in the heart of Gnomeland itself, right in the middle of Zurich, up to the neck in Kreditanstalts and Bundesbanks and les Banques de Quelquechose.

Exciting, more particularly, as we were about to dicker with a Gnome in person, actually a goodlooking, dark haired young man, somewhat above average height. Four languages, of course, and charming in all of them.

Not to go into the contest entirely unarmed we had incorporated into our team an expert handler, manager or agent—an elegant lady stockbroker, who lives near us in the South of France.

Instantly, the two ladies rejected the Gnome's offer of a bottle of champagne by way of an aperitif, so firmly, indeed, that I couldn't squeeze even a dry Martini out of the situation for myself. It was clear—as if it hadn't been clear before—that we were on business bent, a large part of the business being to introduce our handler, manager or agent to the Gnome, and from what we hoped would be their mutual understanding to do ourselves a bit of good.

A short silence fell after we'd ordered lunch. To fill it, I said, 'So this is Zurich, the Home of the Money Wrench,' and got Madame's heel along my anklebone by way of commendation.

A moment later I found there was no need to keep the party going with the lighter kind of conversation because our lady aide and the Gnome had serious work to do: i.e. to find out and to lay bare the metal of the other.

She: ' A friend of mine in New York, J. Strangler Stockbaum, was talking quite interestingly about United Electronic Plumbing, Inc. Spot forward reversible debentures, of course.'

He: (having nodded, once, at the mention of J. Strangler Stockbaum) 'Vielleicht, aber on ne sait jamais about the chances of consolidatory refunding.'

Me: (an aside to Madame) 'Perhaps, but one never knows about the chances of—'

Madame: No words, but a look that would have corroded concrete.

After that I let the experts continue without further assistance, because all at once I felt entirely at home, in Zurich, in this discreetly opulent restaurant, because it was exactly the same as the pub beside Harold's Cross dog-track in Dublin.

The same exploratory feelers: 'Your man said that John-Joe Kelly might be up with a good one from Clonmel for the hurdle.' And the same counters, aimed at further elucidation: 'Maybe so, but th'other fella tole me that the Tralee boys, if the price was right, was thinkin' of havin' a go with the brindle.'

All the Kreditanstalts and the Bundesbanks and the Banques de Quelquechose—there was no mystery about them. They were engaged, one and all, in the holy process of trying to find a nice, long-priced winner.

When lunch was over I asked Madame, 'How did we do?'

'I did fine,' she said. 'He's going to send me a huge box of lovely Swiss chocolates.'

'Is that all? Big deal.'

'Ah,' she said, 'but he wanted to know if I'd rather have white or dark truffles in the centres.'

I suppose I'll find out, sometime, what really happened.

Magnificent Mumbles

THE most extraordinary fantasy occurred to me the other day. It might have been provoked by a surfeit of peaches, which are now approaching the end of their season and can be bought by the dozen at the equivalent of about a penny each.

It was very hot, but I had arranged myself in such a way that the bare top of the head was sheltered by a branch of the orange tree, while the rest of the machinery was in the full glare of the sun. A certain langour developed, with this bowl of chilled peaches by my side. Indeed, it was difficult to keep the eyes open. Several times I was only just rescued from unconsciousness by taking another peach, biting into it, and allowing the cool juice to run down the chest. The tiny frisson thus provided served to keep the mind alive, although admittedly not up to the full capacity of this well-known power house.

Suddenly, then, and without warning, it sprang into full production. A thought had descended upon me, with the force of a double-decker bus falling over a cliff.

What would it be like, I thought, to take a holiday from here?

I actually groaned aloud at the prospect in store. Not, that is, at the prospect of taking a holiday, but at the prospect of thinking about so doing, with all the fearful back-to-front work involved.

People take holidays because they want to get away from the roar of traffic as it splashes in heavy rain down their suburban road at the end of which a battery of pneumatic drills has been digging a huge hole for the last month. They want to go to the sea or the mountains or the lakes. They want to find sunshine and an atmosphere so different, say, to that of their native Liverpool, Leeds or Manchester that they will feel reborn for a whole three weeks.

Cheese-coloured faces and limbs will become a little brown. There will be the pleasure of speaking a few words of a foreign language and being understood. No struggling with public transport on the way to the office. No waiting for public houses to open. No sodden walks with the dog on Sunday mornings because there is nothing whatever else to do. No collars, ties, socks, overcoats or bowler hats. Even, for the truly adventurous, no more television for three glorious, self-sufficient weeks. In place of Eamonn Andrews every evening, just sitting in a café watching the rest of the pleasure-seeking world go by, with something in your hand that isn't a pint of luke-warm bitter.

At the thought of a pint of luke-warm bitter I took another chilled peach—the white kind, which is just that little bit more tart and tasteful than the yellow. I allowed plenty of juice to run down the bare chest, because now the hard part was coming, and I wanted to be able for it.

I put it to myself as bravely and as bluntly as I could.

What kind of holiday can you take when you live in almost continual sunshine in an olive grove in the mountains, which is only twenty minutes away from the beaches and the sea?

I wished I hadn't started it.

Glorious Peebles? Magnificent Mumbles? Delightful Devon? The preparations involved! In place of buying light-weight, colourful shirts and shorts, going down to the cellar and opening up the trunks and making a selection of the heavier vests, the woollier sweaters, the thicker socks, the rubber boots, the mackintoshes and the overcoats, for three weeks careless rapture at Flacton-on-Sea.

Breakfast—8.30. to 9.30. Lunch—1 p.m. to 2 p.m., and look slippy about it because what with the Catering and Wages Act and SET and the licensing laws it's hardly worth our while to feed you at all. Nor is there any call for the pleasurable necessity of speaking a foreign language. You can only listen to your own, emerging endlessly from the man who monopolises the same stool at the morning and evening sessions in the American Bar at your Royal Hotel. You mark his words— unless these damn Labour chaps mend their ways pretty soon and pull their socks up and get their finger out this old country of ours is pretty soon going to be in an even worse mess than it is at the moment and on and on and on. And to get away from it you get out the car, because it's raining, and instantly you're breathalysed by three policemen while six more frisk you for drugs . . . I came to a bold decision. We will *not* be taking a holiday this year.

Rioting Nicely

MAN, does this rioting take it out of you, or does it take it out of you?

After last Sunday's little lot I've been as stiff as a board all week.

It gets me mostly in the bowling arm. One seems to be dispensing fruit and veg in every direction absolutely for *hours*. And then, of course, the weight of stuff we have to carry isn't any too good for the old back.

Let me give you a list of my own equipment, for instance, for working in Grosvenor Square.

First, a bunch of at least twenty-four anti-police horse bananas, with a stout cord threaded through the main stem so that the weapon can be secured across the chest. Then, a large bag of anti-police horse variegated marbles slung on the right hip.

It takes ages, of course, to get this properly adjusted. It's got to be at just the right height for easy marble distribution and be capable also of quick release, in case anything goes wrong. My friend Caroline got an awful bruise on her leg from her bag of marbles. She was running away from some boys who rather disloyally wanted to riot with *her* and she galloped right into the base of the Roosevelt Memorial and practically drove her marbles right through her thigh. The silly thing. If you're going to riot properly you've *got* to have some method of ditching the bag before it ditches you. I thought everyone knew that.

45

Then, there's my dustpan and brush for gathering up the marbles in case there's another horse to do, though of course some people aren't as conscientious as me. Some people just distribute the marbles as soon as they even *see* a police horse and then chuck down all their banana skins and run, whereas the proper rioter stands his ground and *helps* the horse to fall down, because of course it's only after the horse has fallen down that you can fill the fascist policeman's cap with cream cheese, or whatever instrument of protest you may be using.

I carry a mixture of marmalade and suet, myself. I find it makes my point clearer and more lasting and what's more I carry it in a stout tin. My friend Jeremy always uses plastic bags for his various protest fluids, trying to save weight, but last Sunday it didn't save his brand-new mauve corduroy bell-bottomed trousers. It certainly did not.

It was all the fault of those horrid Irishmen. They were having some sort of protest of their own against St. Patrick, or someone like that, and they got mixed up with our lot in Trafalgar Square. Jeremy was just shaping up with a currant bun to let fly at Vanessa—he's never liked her—when one of these horrid Irishmen simply slit Jeremy's bag of anti-American ink and tomato juice and it absolutely gushed out all over his new cords.

Jeremy told them it wasn't fair and they threw him into one of the fountains, so that he had to go home and change and lost at least an hour of valuable rioting time. Actually, he tried to take a taxi back to Trafalgar Square but the cabbie wouldn't go any farther than Hyde Park Corner, so Jeremy protested half a dozen eggs all over the mudguard and the brutal, war-mongering cabbie gave him the most awful kick on the bottom. Not Jeremy's lucky day.

Where was I? Marbles, bananas, filling for fascist policeman's helmets, dustpan and brush—Oh, yes, a large bag of potatoes for protesting through embassy windows.

Some people I know, like Julian and Arabella and those

rather rough boys Ron and Len, people like that seem to prefer stones and large ball-bearings. They say they drive the protest further home, but personally I find if you can get enough speed on the spud you can make your convictions known even through the window of the ambassador's bedroom on the top floor. And of course the best of the spuds is that you might just be taking them home from the supermarket, so that you're not armed with malicious intent and the Fuzz'll have their work cut out to do you. If you're in rioting seriously you've got to keep thinking ahead like that all the time.

Well, I've got to go and limber up now, getting ready for Grosvenor Square again this evening. A couple of pounds of old pears against one of the lesser embassies ought to do it. I only hope it isn't going to rain.

A More Likely Explanation

IT was one of those standing up parties where a lot of people who hadn't seen one another for about a week were gossiping about mutual friends whom they hadn't come across for something like a fortnight—good potential raw material, if only I'd known the name of even one of the victims.

As it was, time was dragging a little. The only thing I could find to do, in fact, was to change over the jackets on a number of books in the shelves beside my elbow, slipping Anthony Powell on to Kingsley Amis, and so on, but it wasn't very rewarding, in that I would not be there to watch the increasing puzzlement on the face of the next reader. Then this rather saturnine looking young man came over to join me. He, too, seemed to be on his own.

'It's rather better,' he said, looking at the book in my hand, 'if you put them on upside down. It just makes for that extra slight delay in discovery, and therefore enrages all the more.'

'Thanks,' I said. 'I'll try to remember that for next time.'

He leant languidly against the wall. 'I thought he arranged the rain rather well,' he said. 'Though it was probably rather a bore for the alleged Jackie. He had half a dozen of these giant tankers of his on the other side of the island, spraying millions of gallons of fresh water into the air so that it would fall on the alleged ceremony, bringing it luck in the immemorial Greek tradition.'

'Very well done indeed.' I said, not to be outfaced. 'I believe that the operation, excluding of course the cost of the water, set him back nearly half a million. Crews wages and so on.'

'Fuel, and that,' he said.

I found I was compelled to ask the next question. 'This 'alleged' Jackie,' I said. 'You mean it wasn't really—her?'

He took a coutious look around the crowded room, then lowered his voice. 'Of course not, old man,' he said. 'It was a stripper called Mitzy Dawn. Absolute dead ringer for Jackie, including that deliciously breathy little voice. Tricky Dicky arranged the whole thing.'

'Dick Nixon?'

'Who else? He got the wind up about Humpty Dumpty Humphrey coming with a late run, so he got this brilliant idea about undermining the Democratic party for ever. I mean, imagine Jackie Kennedy—the Democrats' shining princess—abandoning the ship just a couple of weeks before the presidential election and nipping off and marrying Nick the Grik.'

'Nick? But she—or, rather, Mitzy Dawn—called him 'Tellis'.'

'My dear, good man,' he said patiently. 'You do not imagine for one single second, do you, that the bridegroom was Onassis?'

He had to laugh, and very well done it was, too.

'Onassis,' he said, 'was aboard one of the tankers, supervising the rain production, chuckling his head off, while Nick and Mitzy Dawn got spliced. Nick by the way,' he added, 'used to sell guide books at Delphi. Not even Madame Callas could have told the two of them apart.'

'If you don't mind my asking,' I said after a moment, 'where was Jackie while all this was going on?'

'With the Duke of Windsor,' he said. 'They were hoping to get married quietly while the heat was on the other two but the divorce didn't come through in time.'

'Divorce!' I said. 'The Windsors! But what's she going to—'

'Marry Tricky Dicky, of course,' he said, 'after Pat's been paid off. She always wanted to be Queen but First Lady is a lot better than nothing. You'll find she'll be a marvellous hand at the job.'

I thought for a fairly long moment. 'What about Onassis himself?' I asked in the end. 'He seems to be left out in the cold.'

'Not him.' the young man said. 'After Dicky gets to be President he's going to give him the whole State of California— gambling rights, nightclubs—the lot. Onassis has always been put out over losing Monte Carlo. California will do him fine.'

'Well,' I said, 'I don't know. It all seems to be just a little— improbable.'

He looked me straight in the eye. 'Can you,' he said, 'think of a more likely explanation for the marriage?'

Bonjour, Redeemer

A BIT batchelory last Sunday morning, with Madame taking a short holiday in Paris and the rain, thinking it was Wigan, absolutely thundering down on the Côte d'Azur, as it had been doing for most of the night.

The general mood in a lowish gear as I came a little arthritically down the stairs about 9 a.m., carrying the breakfast tray, still in pyjamas and dressing-gown and still unshaven with the pillow-pressed hair standing up all round the head. Definitely fusty and crusty and best left alone, at least for a couple of hours.

I was washing up the breakfast when someone knocked on the front door. On a morning like this, and at this hour, I knew it could only be André from next door, wanting to use the telephone. Still carrying a sodden dishcloth I opened up for him and found it wasn't him at all but three other people. A young woman in an oatmeal suit with a bit of fur round the neck, standing under an umbrella: a small, apparently female child in a hooded plastic mackintosh; and a neat little man, about jockey size, with crinkly grey hair, also plastic macked but without a hood.

I made this instantaneous and detailed invertory because they were so unlike what I'd been expecting. The young woman seemed equally surprised, as she took in my own appearance.

Then she said, in the normal French phrase, 'Please forgive me if I derange you.'

On this occasion it seemed to have a special force, but I said, 'Not at all, Madame.' I tried to do something about my hair. The rain swooshed down on the three visitors.

The young woman, encumbered by the umbrella, succeeded in opening her bag, at the same time explaining the purpose of their call in very rapid, nervous French. I caught almost nothing of it. I'd been reading Henry James over breakfast in bed and my mind was steeped in the Old Master's dense, English verbiage.

From her handbag the young woman produced a thin, papery looking little magazine. On the front of it, in large letters appeared the words, 'Reveillez-vous! L'Eglise Catholique de Notre Temps.'

Until now I'd had a faint hope that they were looking for directions. A lot of new houses have been built around here and at weekends people come up from Nice to visit relatives, and lose their way. Now I realised they must be on a collecting drive.

The young woman offered me the magazine, protecting it with her umbrella. I took it from her very loosely, showing it wasn't going to stick. I spread on a polite but regretful smile. 'I'm sorry, Madame,' I said, 'but it's not for me. You see, I'm Protestant.'

For the first time the small man spoke. In fact, they both spoke together, eagerly. 'That doesn't matter!' they cried. 'It doesn't matter at all!' The small man added, for further emphasis, 'It's all the same thing.' They wanted passionately to set my mind at rest upon this score. They couldn't bear the thought that ignorance might deprive me of their message.

The rain thundered down upon them. They looked at me with bright, expectant faces. The child was sucking its thumb. 'Won't you come in,' I said, nailed right up against the wall.

We arranged ourselves in the sittingroom, the three of them in a line on the sofa, while I faced them on the stool in the window. I found I was still carrying the wet dishcloth. I put it in my lap, where it began to soak through immediately.

The young woman began briskly, as though selling a desirable commodity, 'Do you believe in the Kingdom of God?'

If I'd been fully dressed, with my hair combed, and speaking English, and furthermore if it had been late at night, I probably would have been more than ready for theological discussion. As it was I said, 'Mais—oui.'

That did it. For nearly an hour, with frequent readings from the Books of Timothy and Luke, she proved that the world was at its final crossroads, with Youth battling against Age and the ultimate anguish every-where. Then, for five francs, she sold me a year's subscription to the magazine. They left, beaming with joy. I looked at the receipt and suffered a fearful shock.

Madame returned from Paris next day. 'I've had some visitors,' I said. 'Jehovah's Witnesses. They're coming to see you, too.'

Madame looked at the receipt. She spoke one word. 'Right', she said.

I'm delighted she's back.

Also, has Tamara's Cat had Kittens?

PAINT the six french windows in the house beside the pool. This will involve going to Grasse to buy twelve rolls of sticky paper, because if I don't the paint will be all over the glass and it never comes off with a razor blade.

While I'm there buy 10 (20? 30? 40?) kilos of grass seed to re-sow the trenches which were dug for the automatic watering system. Not that it looks as if there's going to be any need, ever again, for automatic watering after the torrential deluges of the last few days. It's been so wet, indeed, that the filled-in soil in the trenches has sunk well below the level of the existing grass, so that at least three lorry loads of earth are required for topping up. But at any moment now the builder, who has been with us since April 1st., will have finished *his* work, and will have repaired the fence, so that there's going to be no method of getting the lorries on to the property.

Think about all that another time.

Buy six stout laths to form cross-pieces in the wine cellar to prevent the third and all subsequent rows of bottles sliding out on to the floor.

But this will invlove finding some preparation which will dissolve the rust which has welded the drill to the chuck in the electric whizzer, because unfortunately the stuck drill is for masonry and incredibly enough it will not bore a hole in wood. And holes have to be bored in the laths because the uprights are made of ancient oak into which screws will not go so that enormous nails will have to be bashed in instead. But if the

54

enormous nails are bashed into the laths they will split, so they have to have holes bored in them.

Buy some enormous nails.

Buy 500 sheets of sandpaper, just in case Madame has her way and a professional housepainter is called in to finish the job so nobly and almost efficiently started so long ago by my amateur self. An awful lot of sandpaper is needed for rubbing down—or even entirely off—the first coat of paint on the french windows, because I have to allow that it is just a little bobbly in places and if this professional is called in before I can smooth down the bad bits he'll want to burn it *all* off, in the interests of preserving his reputation, and he'll be with us for Christmas.

Have another shot at de-rusting the electric drill because if I could get the chuck off I could put on the sanding device and sand the windows mechanically and that would certainly put this flash professional in his place—if he ever turned up.

Have a look at the metal-framed windows, still unpainted, for the bathroom and the potting shed, and see if it's possible to spread putty fairly evenly between the glass and the catch. This was not done by the flash professional glazier, probably on the grounds that it was impossible, but someone's got to do it and no one else is around.

It's a process that will certainly spread a lot of putty all over the glass, because the gap between the catch, which is three inches long, and the glass is about $\frac{1}{2}$ inch wide. The handle of a teaspoon?

Look up the French word for putty, and armour myself so far as possible with other words to describe its texture, colour and consistency in case the man with the long black moustache in the paint shop asks me, as he always does, exactly what kind of putty I'm looking for.

Drive up the road to Tamara Desni's bar and see if her cat has as yet had her kittens and choose the one that we want before she gives it away, keeping the one we don't. Make absolutely certain that these negotiations are confined to one hour at the very outside, because the last time we had lunch there we left at 9.35 p.m. and it's too tiring.

Go and find a piece of marble for a top to the sewing-machine base that Ruby kindly gave us and either before or after doing so see if it's possible to detach the treadle mechanism because if not someone is going to get their bare feet treadled every time they go anywhere near it.

De-rust the electric drill because the bit that's in it is too small for the holes that will have to be bored in the marble to fasten it to the sewing-machine base.

Ask someone if it's possible to bore holes in marble without splitting it in every direction and then cut the grass and buy some long poles to support the six new mimosa trees which have shot up like beanstalks after the rain and most of all have this memo photostatted a thousand times to give to the people who ask, 'But what do you *do* all day down there?'

Gullible and the Baedeker Kid

As we heard the clunk which indicates either fatal fracture of the wing roots or the safe folding in of the under-carriage the man sitting next to me clicked a stopwatch.

'Eighteen point five secs,' he said, in an official tone, 'with a seventy-two per cent payload.' He relaxed, allowing himself a small chuckle. 'Not bad,' he said. 'But then Doggy Moorehouse is always one for pressing in.'

He spent a few moments polishing the stopwatch with his handkerchief, then lowered it into a small, maroon coloured pouch which he subsequently returned to his waistcoat pocket.

Out of the whole 72% payload—if by that he meant the passengers—it seemed probable that he was the only one who was wearing a waistcoat. A waistcoat, with its slightly old-fashioned ambience, suited him. It went with his thin, National Health glasses, the red, button nose and the gingery, tufty hair.

I'd noticed him briefly before take-off, when the hostess was handing round the boiled sweets, because he'd filled his pockets with a couple of handfuls. He was one of those men, in the middle forties, who preserve both the nature and the looks of an earnest and inventive schoolboy of about eleven.

'Oh, yes,' he said. 'Old Doggy presses on, all right. Clocked two thousand four hundred and twenty-three flying hours. I've been with him to Montreal, Rio, Philadelphia and Carracas. He's not married, you know.'

I had the feeling that quite a lot of indigestible, not to say

57

unrequired, information was being handed out. 'Obviously,' I said, 'he hasn't time.'

My companion turned towards me, efficiently pushing the glasses up on his nose. 'Ah,' he said warmly, 'but that's where you're wrong. There,' he said, nailing it down, 'you're very very wrong indeed. Doggy just doesn't go in for that sort of thing.'

'Even in Carracas?'

For a moment he looked shocked. 'Surely,' he said, 'you must be joking. The stop-over in Carracas is only fifty-eight minutes, and he's only there once every fourth week.'

'I'm sorry.'

'That's all right,' he said comfortably, sitting back. 'It's easy to make a mistake if you're not in possession of all the facts. Would you like to see my log-book?'

I'd thought I'd been dismissed as someone too light-minded to support even the passive end of a conversation, so that I fatally taken aback by this unexpected enquiry.

'What sort of log-book?'

It was his turn to be surprised. 'Flight log, of course,' he said. He reached into a livid-yellow airline bag, of a kind I'd never seen before, and brought out a thick, chunky volume bound in grey serviceable linen. 'It's all in here, you know.'

Plain curiosity caused me to reach out my hand. He almost snatched the book away. 'Just a sec, old chap,' he said reprovingly. 'I've got to fill in m'details, don't I, before I forget?'

He selected a Biro pen from a battery of six in one of the pockets of his waistcoat. For the next couple of minutes he filled in various columns with extreme concentration. He put the pen back again. 'There,' he said, 'that's all ticketyboo.' He looked at me suddenly over the top of his glasses. 'Would you like a sweetie?' he said. The idea had just come to him.

'No, thank you.'

He produced a handful and made a careful selection. 'Barley shug,' he said. 'Yum.' He worked on the barley sugar, moistening it and distributing it sufficiently widely in his mouth to permit him to speak. 'Now', he said. 'Here we go.'

He put the thick book on his lap and opened it at the first page. He began to read, surprisingly loudly, 'May tenth, nineteen hundred and thirty-two, Croydon Airport, destination Paris, advertised time of departure ten-fifteen, actual time of departure ten-twenty-one, Captain B. Smith, weather fair, ceiling eight thousand feet, wind nor-nor-west.'

It seemed to go on for ever, but we reached Paris in the end. He finished his barley sugar at the same time and slotted in another one. 'That,' he said, around it, 'was Flight Number One. Now, Flight Number Two. May fourteenth, nineteen hundred and thirty-two, Paris Le Bourget—'

I had to stop it. 'You mean you've kept a record of every flight you've ever made?'

'Of course!'

'Are you in the airline business, then?'

'No, no,' he said. 'I'm a journalist. Travel writer. You may have seen some of my stuff.' Rapidly, he reeled off the names of seven provincial newspapers, and three technical magazines.

'I'm afraid,' I said, 'I don't often get to read too many of those.'

'No?' he said, interested in the revelation. 'Well, I've got quite a lot of them here.' He reached into the yellow bag again and produced a thick bundle of cuttings, of column length. 'This one might interest you,' he said.

It was headed, 'Colombia's Capital—Booming Bogota.'

'You mean you've been there already!' I exclaimed.

'A couple of months ago.'

'And now you're going to write about it all over again?'

He popped in another sweetie. 'Well,' he said, not in the least defensively, 'it's a free trip, isn't it?' He chewed in the same preparatory fashion as before. 'Now,' he said, passing a finger down the column, 'there are some figures here about Colombian State Railways that will probably interest you—'

'I'm frightfully sorry,' I said, 'but I've simply got to go to the lavatory. I'll be back in a tick—a sec—I'll be back.'

I stood in the lavatory for some time, in a state almost of shock. 'Jesus Christ,' I said to myself. 'I don't believe it.'

The proposition was indeed unbelievable. Here was a marvellous free trip to South America, to Bogota, one of the highest cities in the world. Tropical heat, Inca remnants, bandits and the week-end in Cartagena, on the Caribbean, and this man with his flight log-book and National Health glasses and his limitless stream of corroding information would be right beside me, every inch of the way. It simply wasn't fair.

On the way back to my seat I stopped to have a word with Joe. 'Who's this character I've drawn?' I asked him. 'The one who's sitting beside me. The world-wide mine of information.'

'You,' he said, 'have got Mrs. Baedeker's Little Boy, and all of us here wish you the very very best of luck.'

Some of the other newspapermen raised their glasses, in solemn toast.

'I'm going to sit here,' I said.

Joe said, very earnestly, 'He'll come and lean against your seat and you'll get it all down the back of your neck. If I were you I'd just face it bravely.'

Mrs. Baedeker's Little Boy was glad to see me back. He was even concerned for my welfare. 'If you've got a touch of gyppy tumtum,' he said, 'I've got the very thing for it here. Two per cent opium extract, two and a half per cent magnesium sulphate—'

'I'm all right,' I said. 'I—I've just got rather a lot of reading to do.'

In fact, it was a paperback lump of Harold Robbins, which I'd bought in a hurry at London Airport. To stop Mrs. B's L.B. opening his mouth I said, 'Trying to turn it into a film script. Wants a lot of cencentration.'

He nodded understandingly. He seemed to be impressed.

Perhaps writing for the screen was the only thing in the whole world he didn't know everything about.

Later that evening we prepared to land at Madrid. He took an unusual amount of care with his seat-belt. 'Doggy,' he volunteered, 'doesn't much like this one. You've got your mountains all round the airfield perimeter, you see, rising in places to anything up to—oh—two thousand feet. No, wait. I'll just check that.' He was still audibly checking it, from his log-book, when we made a slightly bumpy landing.

He looked at me, humourously, over the top of his awful glasses. 'See?' he said.

It was a long night, as we flew across the Atlantic to Puerto Rica. During it, I learnt everything I wanted to know about the railway system of Colombia. At one point my companion broke off to ask if he was boring me. 'You see,' he admitted with rueful honesty, 'I'm really a railway man, myself. In fact, I'm hoping I can get out of flying down to Cartagena. Much rather go by train—'

I had a vision of a rickety old Victorian railway carriage enveloped in sheets of red flame, while bandits poured thousands of rounds of machine-gun fire into the ruins, but it didn't work. It was too good to be true.

In the morning, at Puerto Rica Airport, there was a calypso band and a lot of lovely mulattos and octoroons around, but I got the figures for American aid plus those for Puerto Rican emigration during the past fiscal year.

During the flight over the sparkling Caribbean I had to endure the agony, being the lesser one, of reading Harold Robbins for the second time. The airfield at Carracas, on the coast of Venezuela, seemed to be alive with wild and rapacious dogs, so that unlike everyone else Mrs. Baedeker's Little Boy and I sat in the plane, while we digested recent figures for the incidence of rabies in South America, coupled with those for foot and mouth disease.

It was very hot indeed. I sat there beside my companion,

listening, screaming silently but with diminishing vitality, 'SHUT UP! GO AWAY! Burst. Explode . . .' He had changed into a shirt covered with brightly coloured bananas purchased not—as one might have supposed—in the tropics, but in Sauchiehall street, Glasgow. He wanted my assurance—in a sudden parenthesis—that it wasn't a bit too swish. I gave it to him, with all the earnestness at my command. He was relieved. At heart, he explained, he was really a bit of a buccaneer, inclined—in hot countries—to cut a bit of a dash. Sometimes he went just a little too far, but—thank goodness—he realised this fault.

'Know thyself, eh?' he said.

I nodded in sombre agreement. Then we went back to rabies and foot and mouth.

The first two days in Bogota were marvellous. An extraordinary city of concrete skyscrapers, with bulldozers in the rutted streets, and the empty ground floors of the office blocks occupied by whole families of Indians, cooking on camp fires. A city nine thousand feet in the air, with gangs of ferocious bandits alleged to be in the very suburbs. And the best of it was that Mrs. Baedeker's Little Boy seemed to be extremely busy with something else. I saw him only in the evenings, dining at a small table by himself, surrounded by piles of books of reference. Once, I caught his eye and he rather gaily raised a thumb in the air. I couldn't imagine what he meant.

When I found out, it was like being sentenced to death. What he'd been doing, for the last two days, was working out the most interesting and varied method of getting from Bogota to Cartagena by train. It hadn't been easy, because the Station Master spoke only Spanish, and all the time-tables were naturally in the same language, but he'd done it in the end.

'The only snag is,' he said, 'we'll have to leave fairly early tomorrow morning—o-five-thirty-four, in fact.' Then he added a reassuring detail.' But it only takes twenty-four hours.'

I said, slowly and carefully, 'I am not going with you in the train. I am going in the aeroplane.'

He couldn't believe it. 'But you *said*,' he cried repeatedly, 'you *said* you were a railway chap too!' He was outraged, unable to credit such duplicity.

I shouted at him hysterically, 'I hate railways! I can't stand the bloody things. I'm going in the plane, d'you understand? *I'm going in the plane!*'

Without warning of any kind he suddenly said, almost humbly, 'Have you got a sweetie?'

'Of course I haven't got a sweetie!' I bawled. 'Jesus Christ, I've only been here for a couple of days and I don't speak a sodding word of Spanish. How the hell do you think I could have got myself a sweetie?'

'I meant a barley shug,' he said in small voice.

It was a peace offering. He'd capitulated. He was prepared to go in the plane.

I felt sorry for him. 'You didn't pay for those train tickets, I hope?'

'Goodness me, no,' he said, 'I wangled them out of the airline.' Hope surged back again. 'You're sure you won't come in the train?' he said. 'It won't cost you anything and I'm sure you'll find it most interesting. They've got a 1923 Series 2B loco with a four-o-four bogie on the—'

'Stuff it up your jumper!' I shouted at him, almost out of my mind.

A roasting hot trade wind thundered through the open balconies of the hotel in Cartagena. The sea-water pool was bliss. The old city itself was like a film-set, a maze of Moorish alleyways, stately French Empire buildings and everywhere people of every imaginable colour lounging in the heat, chatting in half a dozen languages—the very essence of the Caribbean. Morgan and his pirates were just around the corner—unlike Mrs. Baedeker's Little Boy.

Him I saw only once, in the first three days. The buccaneering side of his nature had taken hold, irresistibly in these surroundings. To the banana shirt he had now added a truly enormous, high-crowned, floppy hat, of red and black woven straw. His face was scarcely visible beneath it but as he passed me in the hotel corridor I caught a flash, in the National Health glasses, of the coolest imaginable disdain.

On the last day we went on a conducted tour, with a Spanish interpreter, of the old fortress that guards the harbour. As our guide, in broken English, told us something of its history Mrs. Baedeker's Little Boy—a figure of fantasy in his straw hat—lounged about on the outskirts, denying the accuracy of almost every single fact given us by the guide. His voice was no more than a low monotone, but it was interspersed from time to time with a loud and contemptuous clicking of the tongue that had a great effect.

On the topmost pinnacle of the old fort was a new-looking concrete pillbox, with a complicated radio aerial sticking out of its roof. As we came closer we could hear the rapid, metallic stuttering of the Morse Code. I was aware of Mrs. Baedeker's Little Boy standing beside me. Ludicrously, he had pushed his vast straw hat to one side, and was listening intently with his right ear.

'Yes,' he said, in his official tone. 'I thought so. The Mauretania—about 180 miles due south of us, making for—Miami.'

He returned the hat to its original position.

I seized the interpreter by the arm. 'Even if it's illegal,' I said, 'go into that hut instantly and ask what that last signal was all about.'

It was all about the Mauretania, 180 miles due south of Cartagena, making for Miami—and no mistake about it.

I rounded on Mrs. Baedekker's Little Boy.

'How,' I said, 'did you know that?'

He was surprised. He pushed up his glasses. 'Well,' he said, 'a chap keeps up his Morse, doesn't he?'

Every time he spoke to me, all the way back to London, I said, 'Dit dit dit dah dah dit dot dash bash nit dit dit dah.'

He accepted it with weary resignation, obviously hoping that our paths would not cross again.

Splice Every Brace in Sight

The following extraordinary message has been received by the Sunday Times radio station, which benignly day and night monitors the world. I reproduce the operator's transcript in its entirety.

S.O.S. Mayday. Au secours. Help. And tell all the lads in the Pig and Whistle to have another one on me.

(Operator's note: For several minutes after the reception of this alarm call nothing could be heard from the sender except the sound of singing. The number in question appeared to be that old but popular favourite, 'Nellie Dean'. Transmission was then resumed.)

Cor, that was a near one. Ruddy big lump of a thing—looked like the Queen Elizabeth—damn nearly run me down. Never even saw her. Comes belting at me out of the blue and spilt me drink. You'd think they could be more careful. I'll write to someone about it if I ever get ashore. Although of course I don't know if I'm allowed to.

(Operator's note: At this point I recorded the sound of a cork popping. This was followed by a short period of silence, which was broken eventually by an eructation. Transmission was then resumed.)

Trouble is I've forgotten which race I'm in. I remember lashing out from Plymouth or Southampton or wherever with me Genoa set and me spinnaker—I'm ashamed to say—wrapped round the top of the stick. I thought I was hoisting me mainsail, but it turned out to be the other thing. Probably some damn fool ashore, fiddling with things he didn't understand. Where was I? Yeah. To put it bluntly, in seamanlike fashion, I can't remember if I'm in the transatlantic race or the nonstop-round-the-world lark. You get like that out here in the roaring forties, where it's just man and thirty-seven bottles of malt whisky against the elements.

(Operator's note: At this point the sender sang 'Red Sails in the Sunset' twice. An attempt to sing it a third time was interrupted by a sudden paroxysm of crying. After an interval transmission was resumed.)

Lovely old song. They dont write 'em like that nowadays. How are you, Gracie, me old luv? I'll probably run into you on Capri one of these fine days. In fact, I'll probably run into Capri, head on, at a rate of knots. If, that is, I'm in the Mediterranean and not somewhere south of the Azores. Anyway, wherever I am, and whichever race I'm in, I'm going to break the record. Good old Bill Howell did it on beer. Something like a hundred and twenty eight crates. The French lad, Castelbajac, stuck to the old vino. Eight-two bottles in thirty-two days. But brother, me, I'm loyal to the malt. Thirty-seven bottles under the belt already and plenty more where that lot came from and here's to me sponsors, God bless 'em.

(Operator's note: For some time after this the sender appeared to be moving round his boat. I recorded the sound of a dull thud, followed by the exclamation, 'Cor, me nut!' Transmission was then resumed.)

There's one thing bothering me. I think I'm in the wrong boat. I'm nearly certain I started out in a catamaran. But the thing I'm sitting in now's got three pontoons. A trimaran, if ever I saw one. And the starboard hull's full of mermaids. Decent girls, but a bit silent. I call them Trishy, Dishy and Fishy, but they never speak. Of course, they may well be just a deep-water mirage. A man gets very close to his Maker, out here in the mountainous seas of the North Atlantic. Pacific? Indian Ocean? Straits of Messina? Anyway, wherever it is, a man sees things out here that are veiled from your average desk-bound landlubber. Father Neptune becomes his friend. Nay, his veritable doppelgänger, his familiar . . .Whoa, buddy boy. You're running off at the mouth. Write it down. Save it for your book. Entitled? I've got it! 'Half Seas Over'. A smash at Christina Foyle's literary lunch. I must have a little chat with her ladyship Chichester about what to wear. Perhaps just me spinnaker, caught at the throat with a marlin spike— Cor!

(Operator's note: For several minutes I recorded the sound of someone struggling to regain his feet after falling into a heap of broken bottles. Transmission was then resumed.)

I've done it. Landfall. Land ho! Statue of Liberty—here I come. The winnah!

(Operator's note: It was subsequently established that the sender had confused the Statue of Liberty with Blackpool Tower. Shortly after the break in transmission he struck Blackpool Pier about half way down, fortunately without loss of life. End message.)

Gestures Under Bum

It was Sydney's idea, though he really got it from Ian and Jon who did their thing with the Union Jack so groovily on the roof of Rhodesia House during the weekend.

What really impressed Sydney was the fact that the Fuzz were so kind. 'I mean,' he said, 'they seem to have looked after Ian and Jon for the whole of the seventeen hours they were up there and after they'd finished their thing the Fuzz helped them to get away. I mean, it was beautiful, man. Why don't we do it too?'

Now, I'm as rabid a revolutionary anarchist as anyone I know, but sometimes Sydney makes me nervous. So I asked him, 'Where were you thinking of doing it, man?'

'The roof of Buckingham Palace.'

'With what?'

'A nice big Nazi banner.'

'They'd never let you.'

'We could ask.'

Well, we went along to the Palace and I looked at the ducks in the park while Sydney chatted up the Fuzz on the gate. He came back quite soon, but cool. 'It was beautiful,' he said. 'When I asked the Fuzz if we could get up on the roof with a big Nazi banner he said they were all rather busy inside with Master Charles's investiture, and could we come back later when they'd got things more straightened out?'

'Groovy,' I said, but relieved.

'But it gave me a better idea,' said Sydney. 'Flagwise, man, we're ging to spell it out for them.' And that's how we came to be sitting shortly afterwards on the roof of the block in which Sydney lives, proclaiming our OM to the whole world.

Sidney, you see, had bought a big bundle of International Code flags, the things that ships use for signalling. There's a different flag for every letter of the alphabet, and Sydney had run up O and M. So we sat there under our signal and for about twenty minutes we chanted together, 'OM Om Om OM', just like Alan Ginzburg and all the beautiful people in that Chicago park.

Groovy Zen manifestations, except that it was raining and rather cold. It was Sydney who said, 'Like action-wise we're not making it, man. We've got to kick the passive. We must strike, scream, smash and shock.'

He removed the O of OM and put two different flags in its place.

'What's that?' I said.

'BUM,' said Syndey grimly. 'It's a yelling gesture of defiance against the old, worn-out orders of Church and State.'

Well, we sat under BUM for a long time, letting our defiance flare at the world, until Sydney suddenly said, 'Sod this for a lark,' took down BUM and ran up the most popular word in the English theatre. 'That ought to get 'em,' said Syd.

It did. A few minutes later a young copper joined us on the roof, touched his helmet and said, 'Excuse me, gentlemen, but a retired naval person in that block over there has taken exception to your decorations. They spell out a rather groovy obscenity,' the young copper explained.

'I'm fearfully sorry,' said Sydney. He ran up another four letter word in its place. 'Is that better?'

'I'm sure it is, sir,' said the copper. 'Thank you most frightfully for cooling it.'

But of course Sydney hadn't and this time we got a Superintendant, a right crypto-Fascist jackal about seven feet high.

He said that while it wasn't—yet—an offence within the meaning of the Act to fly that word in the International Code if we didn't take it down we'd be in the nick so quick it'd leave us crosseyed.

Once again Sydney apologised, and ran up four more letters. The Super consulted a notebook, nodded and said, 'Keep it at that, son, if you know what's good for you,' and clumped away.

'It's LOVE,' Sydney told me, 'but not for long.'

That's where he was wrong, because after a couple more hours of flying all kinds of four-letter words, interspersed with LOVE, a soppy looking old copper arrived, chuckled sort of comfortably at the word we were flying and said, 'Mind how you go, gents.'

Now we can't get rid of him. This endless 'Mind how you go' is putting us down. It's Mushville. Sydney's certain it's Mr. Dixon in person.

Seems there's nothing those Fascist Stalinist Cossacks won't do to drag the scene.

Revoltingly Impossible

NOT an unfamiliar situation.

Two nice friends have got to be contacted by telephone within the next five minutes, before they leave their house to come to us, because another and fearful couple have suddenly turned up without warning, and one of the nice friends is suing one of the fearfuls for all kinds of malpractices which, while at the moment not including physical assault, will certainly do so the moment they come together in our sitting room. With inevitable damage to our furniture and fittings.

So, while my wife entertains the fearfuls in the potential arena, I pick up the telephone, with no little eagerness, in the bedroom, and there, by God, are the two old cronies all over again, crossing my line and, indeed, their own, with their endlessly repetitive and unchanging conversation.

'Ow you bin, then, dear?'

'Not too good, dear. Ow you bin, then?'

'Not too good, dear. The wevver, you know.'

'Oh, I know, dear . . .'

I bawl and shout and roar at them to get off the line and hang up and shut up, and the same mindless enquiries go on and on. So I smash down my receiver and try again and they're still at it. They've now got as far as two rival herbal remedies, and I know how long that takes, so I bash the receiver down again and dial 100. By all inconceivable improbability the Exchange is engaged.

By all that's revoltingly impossible the Exchange is engaged at 5.45 in the evening, when business must be over for the day and its executives hurrying home in trains to Chorley Wood. What madness has possessed the housewives of London that they all want to make social calls to one another at 5.45 in the evening, when they should be warming up the telly in preparation for the weather and even worse news of national and world disaster? Or perhaps it is that the herbal remedy ladies have clogged my wires.

I try again. They're now on to poultices, and that's nearly always eleven to twelve minutes, so I call the Exchange again and by the Postmaster General and his Awful Dog it's still engaged. Or perhaps the electronics aren't computing or they've turned back to front?

With scientific precision I replace the receiver, pick it up again, count up to five, and then, dialling powerfully but accurately, driving 1—o—o right home, as it were, I get the engaged signal again.

An unspeakable word rings out. Then I sit there, enveloped in scarlet steam. I'd better head off our second pair of guests by standing outside in the rain. But because of the rain they won't be able to get a taxi. It might be an hour. But within the next ten minutes my wife will certainly have savaged the two fearfuls and there'll be more work for the lawyers than even they had bargained for.

I try the telephone again. The herbal remedies have gone and now it's the mother and daughter who discuss the preparation of home-made baby food—by the hour. Someone has crochetted the telephone wires for miles around. We are just winding two pears through the mincer when my eye passes across the morning paper, which is still lying on the bed.

There is a jaunty question, framed in huge black letters: 'HOW WILL THESE NEW TELEPHONE CHARGES AFFECT ME?' And then smaller, but very cool: 'Your Questions Answered'.

I can't bear to look at it. It's the GPO, flashing its cosiest and hippiest image, and it's enough to turn the stomach, like the prunes now churning through the mincer. I'm about to throw the paper away, when something even worse catches my eye. A slogan, across the bottom of the page. A real button-holer, a little masterpiece of chipper invention:

TALK IS CHEAP WHEN YOU PHONE OFF-PEAK.

But something is wrong with it, something above and beyond its egregiously bouncy character. Of, course—it doesn't rhyme! It ought to be:

TALK IS CHEEK WHEN YOU PHONE OFF-PEAK.

But even the GPO wouldn't be that offensive. What about:

TALK IS CHEAP WHEN YOU PHONE OFF-PEEP:

That's exactly where I am. Off-peep, in between the wires and joined to none of them. Talk is, in fact, *cheep* when you phone off-peep. Cheep-cheep, peep-peep.

I sit there in front of the telephone, cheeping and peeping, ready to slaughter the world.

The Pro-tem Branch of the International Whatsit

'STUDENTS, fellow libertarians, workers, hangers-on, policeman-pushers, yellers, bawlers, tomato-throwers and all fully paid-up members of the East Sheen Maoist ad hoc and pro tem revolutionary branch of the International Federation for Disregarding Traffic Lights During Demos—we are faced today with the most serious crisis we have ever—er—faced.

Speaking to you today as the Deputy Brawls Secretary of the Kafka-Marxist passive-violence North Pudsey Militant Christian Brothers I want to speak to you about—er—faces.'

An old cabbage, shedding yellowing leaves in its flight, speeds from the back of the hall and strikes the speaker on the left ear. This organ has become exposed because he has turned to his right to remonstrate with another member of the platform party, who has been remonstrating with *him*.

The speaker picks up the cabbage and remonstrates it right back to where it came from. He then abandons for a moment his earlier rhetorical style and puts in a personal plea.

'Willya shut up, the lot of you. Can't you see how hard it is? I mean, how can I keep track of what I'm saying when I've got to do all this Maoist ad hoc and pro tem Revolutionary Branch of the International Whatsit first? Why don't I just call you—Brothers!'

Four or five stone of mixed vegetables pour down upon him from the balcony. 'And that,' cries a clear voice, 'comes with

the compliments of the Liberal Executive of the Socialist
League for the Refurbishment of Capitalism according to the
Principles of Lin Yung Bung!'

The speaker disentangles himself from a bunch of unusually
long-stemmed leaks, throws both hands high in the air, calls
out, 'Friends!' in a tremendous voice and then lets the Liberal
Executive in the balcony have a left and a right from a couple
of spuds he's concealed in his fists. Someone starts off the
automatic sprinklers and they soak the assembly into some
semblance of order. Then tea, cocoa and buns are served and
the meeting is reconvened.

'Friends,' says the speaker, 'I want to talk to you today
about faces. All right, Fred, let's have it.'

Fred, Proxy leader of the Anti-Fascist Photographic Cell,
flashes an enormously blown-up press photo on the wall behind the speaker. It is greeted with a storm of cheers and
catcalls and a volley of soft fruit.

The speaker removes most of the fruit with a policeman's
truncheon, and resumes his address.

'This photograph,' he says, 'appeared in the Daily Mail on
the morning after the Great Demo. Here, on the left, in a
beautifully chosen militant-democratic pro Vietcong cloth
cap, is good old Nobby Wrench, Chairman Emeritus of the
crypto-mandarin elite boiler-saboteurs corps. Note his expression. Excellent. Rugged, British-Chinese determination. In the
middle is our beloved leader—your beloved leader—their
beloved leader—the Blessed Tariq Ali, his arm gripped in
loyal brotherhood by good old Nobby.'

A voice bawls out, 'That's not Nobby—that's his bruvver
Chawlie!' It dies away in a strangled gurgle.

'Note the Blessed Turiq's expression,' goes on the speaker.
'Pained. Concerned. Elevated. Almost not of this world,
Zenning away like mad. First class. And gripping his other
arm, on his other side, who should we have but Our Darling
Vanessa. No, no. I beg your pardon. Not Vanessa. She's

resting at the moment. But whoever she is she's a true flower of the Upper Christian-Communist Crust. Caroline? Sue? Priscilla? Whoever she may be she's doing it exactly right, too. Serenely lovely, but bang-on revolution-wise. I say it to you in all seriousness—these three could not have done it better.'

The speaker's voice takes on a more severe note. 'But what do we have in the background?' He scrapes off some elderly avocado pear with the truncheon. 'What do we have in the background but some young nit laughing his head off? Sideways, not even facing our enemies, the Fuzz. It's disgusting. It's simply playing into the hands of the human beasts of the American Marine Corps. *It must not happen again*! Now, in preparation for our next Great Demo, I want you all to prac-tice the Revolutionary Look. You can choose whether you wish to be Serenely Lovely. Pained, Concerned or Elevated, or simply Rugged, Anglo-Asiatic Determined. Clamp it on now, and leave it there till it sticks.'

The assembly falls silent, practising devotedly.

While he's safe, the speaker nips around the corner for a quick pint.

Are You Green and Loaded

THERE's rather more to it, you know, than just being the bloke who holds up that card and asks you if you've got anything to declare and when you say you haven't he demonstrates his faith in your veracity by removing the lining of your handbag.

That's for first-year men. A little crude, a little lacking in psychological nuance, and anyway it brings on the finale too quickly. At one moment all bright convincing smiles and at the next tears, and grovelling apology. It leaves the client no time in which to contemplate the true enormity of his offence—like, say, ten extra fags up the jumper when he's already got the statutory two hundred in his hold-all. It doesn't give him the opportunity to—as we call it—'sweat', and without this 'sweating' he doesn't get sufficiently scarred in the soul, so that on the next trip the bastard is very likely to try it again.

When we mark 'em we like 'em to stay marked, and we have several simple methods of bringing this about.

You may have noticed, when you've landed loaded, and collected your baggage and approached the youngest and gentlest looking Customs officer you can see, that he often says, 'I'm sorry, Madam, but this position is closed. Would you apply to my colleague on the far side of the hall?' And Madam, whose tights are practically bursting at the seams with gold bars or whatever, has to stagger right across the hall, only to find the same young and gentle looking Customs officer, who has nipped round the back, waiting to receive her. He then only has to ask her if she's had a good trip, produce the chalk, think better of it and in no time at all Madam is 'marked' for life.

It's the waiting that does it. You want to give 'em time for hope to ebb and flood, ebb and flood and suddenly then to run out altogether. That's when the tights burst, mate, I can tell you, and that's why we senior officers of Her Majesty's Customs and Excise are so much in favour of the new arrangements at London Airport.

At first sight, of course, they look absolutely daft. You see— allegedly for experimental purposes—we've now got these two Customs channels at Heathrow. We've got the Green Channel, for those with nothing to declare, and the Red Channel for those that have.

The purpose of this is to try to speed up the passenger flow for the Christmas rush and later on for the jumbo jet stampedes.

Honestly, when I first heard about it I had to laugh, and it was good old Victor the Vacuum-Cleaner that set me going. Vic the Vac—and, mate, does he get down into the crevices— has been on the job a long, long time. He really *hates* the customers. It's a pleasure to see him at work, but I'd have to allow that he's a bit limited in the top storey.

'Here,' he said to me, when he heard about this Green Channel caper, 'they'll all be boiling down that pipe. They'll all get *away*!' (For Vic a customer getting away is like having a tooth pulled without benefit of anaesthetic.)

'Don't you believe it, mate,' I told him. 'No one's going to go near that Green Channel. It's going to sound much too like a good, clean, practical and generous idea. They won't trust it an inch. They'll think the Government's having them on again. If they arrive wearing only a pair of bathing trunks with their toothbrush in a paper bag they'll queue up to get into the Red Channel—'

'To declare what?' said Vic.

'Their intention to smuggle in a child's spade after next

year's holiday on the Costa Brava. But then, you see, after queueing for an hour and a half in the Red Channel, with the Green one empty beside them, they're going to begin to think they might just possibly chance the Green one and they back out of the Red one and then lose their nerve all over again, so that we'll have thousands of customers milling around in the body of the hall with no one daring to make a break for either exit. That, mate, is when you'll hear the clatter of the jewellery falling out of the suspender belts. That's when the wrist watches will descend from the mini-skirts. Everyone sweating on the top line. Lovely. Specially when we get the Amber Channel.'

'What's that?'

'For Drug Smugglers Only. The watches, cameras and jewellery boys are going to make a rush for that one, thinking they're clean. We'll collect the loot by the bucketful.'

'I think,' said good old Vic, stars in his vicious little eyes, 'we're going to have the happiest Christmas ever.'

Stick in on Your Pare-Brise

FOR quite a long time it seemed like the very last resort of the beleaguered House of Lords. No less an aristo than the Marquis of Aberdeen, on bended knee before the Woolsack, praying for deliverance from the many enemies of the Upper House.

Or, at least, that's what it looked like in French, when I read the story in a local newspaper.

Whatever the prayer may have sounded like in its original form when translated into French it assumed all the thunderous heroics, the heart-searing tragedies of Racine. You could nearly see the actors up there on the stage, tearing themselves to pieces as they shovelled out the great rolling periods.

Certainly, the Marquis—before the Woolsack—began fortissimo. According to this newspaper he cried:

'*De l'obsession de la conduite, de la vanité de la puissance, de l'infection de la vitesse—Délivre-nous, Seigneur!*'

It was impossible not to try to turn it back into English, with deference to the rhetoric.

'De l'obsession de la conduite—'

The whole of this first verse is obviously a direct attack upon Mr. Harold Wilson, and his current tendency towards mucking about with the Upper House and, indeed, everything else. It can only be translated thus:

'From a certain person's obsession with conducting, guiding, directing, controlling and managing; from the vanity of a man who still thinks he's the boss despite the fact that he's lost

81

nearly every bye-election in sight; from the corruption of the
mind that comes from the belief that if you're slippy enough
about it you can get away with anything—GOOD LORD
DELIVER US!

All clear? Next verse, please.

'*De l'insouciance, de l'indifférence, du démon de l'impatience and de
la tyrannie du temps—Délivre-nous, Seigneur!*'

Many scholars will hold this to be one in the eye for Mr. Roy
Jenkins. Let's have it in English:

'From a mindlessly optimistic and jaunty lack of concern;
from the slapdash quality that comes from knowing if the
worst comes to the worst you can always get a job in the City;
from the impatient devil who continues to believe that Britain
has not only turned the corner but is already miles away up
the straight; and above all else from Socialist tyranny over
modern private enterprise—GOOD LORD DELIVER US!'

Well put, my lord Marquis. Continue:

'*De l'inattention, de la monotonie, des fantasmes de la boisson
et des assoupissements de la fatigue —Délivre-nous, Seigneur*'

Some of us regard this passage as rather lacking both in
good taste and relevance, seeing that it clearly refers to a
former Cabinet Minister who by his retirement from the
public scene can no longer effectively defend himself or, indeed,
effectively do anything. For the record, however, we must
continue with the translation:

'From never really knowing where you were; from constantly
behaving in the same somewhat unfortunate manner; from the
delusion that everyone loved you because we all seemed to be
having such a smashing time together; and, rather specially,
after the roof had fallen in, from the explanation that you
were absolutely exhausted from overwork—GOOD LORD
DELIVER US!'

A little sharp, perhaps, seeing that all is now forgiven and
very very nearly forgotten, but—well—*Ça c'est la vie.*

Last verse:

'*Et preserve-nous Seigneur des lenteurs de l'age and de la rage de vitesse de la jeunesse. Fais que les vehicules deviennent des instruments de bonheur pour ceux qui les utilisent.*'

Rather more general in tone from the previous ones:

'And save us, good Lord, from the interminable squabbling between ancient dons and adolescent students—and for God's sake do something about the traffic.'

The last sentence, of course, provides the clue to the true nature of the Marquis's prayer, taken in conjunction with his desire that his fellow peers would stick it on their 'pare-brise'.

At first I thought this meant 'escutcheon,' until the dictionary provided the correct meaning.

A *pare-brise* is a windscreen.

The Importance of Being Starkers

My dear—I simply cannot tell you. But absolutely shattering. Simply Endsville. You won't believe a word of it, but I swear to you it's true.

You know this ghastly charity thing we have for the village? It used to have something to do with the harvest, or something like that, except that no one's harvested anything round here for years, except a fortune for selling their land to people like us, but anyway this ghastly Amateur Dramatic Society does a bit every year at about this time. Usually *The Ghost Train* because we all know it and can swan through without much bother.

But then this new young couple arrived—rather swish with a foreign sports car and apparently wads of money. They seemed to know all about the theatre and television and that kind of thing so it was really rather a relief to let them take over the Dramatic Society, even if they wanted us to do *The Importance of Being Earnest*.

Actually, we'd read it once at a kind of rehearsal, but the plot was so muddly no one could really understand it, so we did *The Ghost Train* again, that year.

I must say, though, these young people, Tarquin and Fenella Brackett, were quite clever at it and explained it was a kind of satire on society or something, and most of us got a fairly good hang of our parts. After all, that bossy Mrs. Dickson-Drayne only had to be herself as Lady Bracknell and the Vicar simply chewed up Canon Chasuble. Tarquin and Fenella, of

of course played Algernon and Gwendolen and I was Miss
Prism, which as it turned out was lucky because I didn't come
on until the beginning of Act Two.

Well, my dear—let me tell you what happened. I know you
won't believe it, but it's true-true-true.

It started all right with all that rather boring talk between
Jack and Algernon about cucumber sandwiches, Actually I
helped it along a bit by standing at the back of the audience
and laughing at fairly regular intervals, encouraging the
rustics to do likewise. Funnily enough, some of them laughed
when I didn't, but they were probably only trying to please.

Then, just as Jack and Algernon began talking about that
difficult Bunbury business, I became aware of what I can
only describe as a disturbance in the wings. I could hear a lot
of smothered giggling from Fenella Brackett—that girl's
always giggling at something, one never quite knows what. And,
at the same time, I could hear the hoarse voice of Sybil
Dickson-Drayne whispering in a kind of frightened way, 'No,
no—certainly not!'

I simply couldn't imagine what was going on. I mean, the
Dickson-Drayne woman hasn't been frightened of anything
for absolutely years. It sounded as if Fenella was trying to make
her do something she didn't want to do—and then a second
later as if she'd decided to do it. Anyway, I heard her give the
most extraordinary kind of laugh—more of a sort of a fat chuckle
really. And then they came on. My dear—STARK NAKED!
And Mrs. Dickson-Drayne said, 'Good afternoon, dear Algernon. I hope you are behaving very well.'

There wasn't a single sound from the audience. They just
sat there, paralysed, until Tarquin—absolutely coolly—said to
Fenella, who was wearing only a pair of court shoes, 'Dear me,
you are smart.' Then some horrible boy in the audience let
out a piercing whistle, so no one could hear the rest of the
play until Fenella said, with a positive leer, 'Oh, I do hope I'm

not quite perfect. It would leave no room for developments, and I intend to develop in many directions.'

With that, she sort of bulged, practically all over the stage. My dear, pandemonium. Some of the nastier old women holding their sides, and the dirty old men clapping like mad and all those horrible boys whistling their heads off. You couldn't hear a word until suddenly the Dickson-Drayne said, 'I'm sure the programme will be delightful, after a few expurgations,' and rose to leave the stage.

I cannot tell you! I must! Before she could get off the Vicar, as Canon Chasuble, came bounding on, wearing only a fig-leaf and nothing at all behind. Of course, this was absolutely unfair because really he wasn't on until after me in the next act, and some sort of madness seized me and I rushed right through the audience tearing off all my things and I was only wearing my bra when I reached the stage and the Vicar unhooked it for me, saying, 'Allow me, madam.' which was rotten of him because it wasn't in the play at all. But then none of the rest of it was either.

Something very queer seems to be happening to the Theatre these days.

Soaking Up to the National Average

THE British Government's statistical bureau has just discovered, I note, that France has once again broken the world's record for alcohol consumption. Every French person over the age of 16 last year consumed 28 litres of pure spirit, an improvement of 2 litres over their previous best.

These dreadful figures are scarcely surprising if one examines the fourteen to fifteen hours of brazen orgy that disgrace the bar in our own village here every day.

It opens at about 8 o'clock every morning for the sale of beers, wines, spirits, picture postcards, torch batteries, cigarettes, lighters, magazines, newspapers and quite often spaghetti with tomato suace.

The day's soaking begins with the elderly wife of the elderly proprietor sweeping the pavement and much of the road outside.

The proprietor, for reasons of economy and hygiene, is in favour of this activity, but it is opposed by the lady's sons, not because of any impairment of their mother's dignity, but because they feel that the pavement and the road are not too bad the way they are.

While all this sweeping is going on it is not possible to obtain a drink because the two sons like to watch their mother doing it. They lean against the wall pointing out to her that very little dust is being gathered, and that in any case the traffic of the day will certainly put it all back again.

At this time the proprietor himself may be behind the counter,

but he will be sitting down and as he is on the short side only
the bobble on top of his woolly cap will be visible. Even if he
is there, however, there is no point in approaching him because
he will serve only three other elderly gentlemen, who are
close personal friends.

When the sweeping is over the whole family disappears into
the back part of the house, presumably for breakfast, until
about 10 a.m. During this break the two sons take it in turn to
shout, 'Oui, j'arrive!' in response to the roars of the drink-
hungry customers, but they have never been known to do so.

The day's drinking really begins after 10 a.m., but one
wants to pick a brilliantly sunny day for it, so that you can
see what you're doing. The bar is long and narrow, with no
windows, and the short counter is at the end away from the
door. The proprietor does not permit the use of the single
bulb, which illuminates the place, until after nightfall, and
then only if customers are actually present. Frankly, it's a
bit gloomy for a booze-up until the sun rises above the house
opposite and a thin ray or two reaches the back of the room,
but in any case there's not much point in looking for a drink
before eleven, because in the preceding hour everyone comes
in to collect their copy of *Nice Matin* and they stand reading it
in front of the counter. A number of small children, buying
sweets, add to the barricade between the rumhounds and the
bottles, so that it's really better to wait a while until the
crowd has cleared, except that of course at midday all the
road-menders and the builders' labourers and the plumbers
come in for their lunch-break.

Many of them are Algerians and they bring their own food—
a loaf of bread three feet long per man—and they sit in silence
at four of the six tables, eating it. It's not very easy to get past
them, or their loaves, to the bar, which at this time of day is
often graced by two motorcycle cops, armed to the teeth and
wearing white crash helmets and cavernous black goggles.
They take a glass of pink, non-alcoholic sirop and before

leaving shake hands with eight or nine civilian acquaintances, who have been waiting for some time for them to go.

It's round about now that the first spirits of the day are served, two microscopic glasses of pastis topped up with bottled tomate juice—an appalling combination known simply as a 'Tomate'. The recipients are two old gentlemen who, I must admit, look as if they'd had a couple in the past. I rather try to avoid them, because they always buy me one each and then I have to buy them one and it puts me off my lunch.

This, in fact, is the peak sozzling point of the day, because after lunch the proprietor and his family tend to go to bed, leaving a small boy to look after the shop. He brushes up the crumbs left by the Algerians and is too busy to serve a drink. It's also difficult to get one after six because then television breaks out—mostly the late Gary Cooper, speaking French—and everyone is too busy watching him to pour even a *tomate*.

After that it requires some pretty sterling work at home to keep anywhere near the national average of 28 litres of pure spirit per man per year.

Four Thousand Here Tonight

I HAD a grand new job yesterday.

I was gev this little notebook, y'see, and a sort of a class of a pen thing to go with it, and all I had to do was to go down the street and knock on a few doors and ask the fella that opened it how many scoops he'd take on an ordinary sort of a day.

Of course you're not slow eether. In fack, you're ahead of me. What happens, you'd like to know, if the old woman herself opens the door, with the Pioneer badge up and all, and Father O'Fiddle in person taking a cup of tea with her in the front parlour? It'd be a hard man, after your way of thinking, that'd have the neck to ask herself how many jars she'd have after the black fella'd gone. You don't hafta tell *me*.

Fill it up again, Mick. Sure we might as well be crucified as the way we are. Where was I?

Me grand new job. Well, y'see, I was taking me ease at the Shelbourne Park dogs when this low-sized class of a fella with a soft hat and an old mackintosh on him comes up to me after the third race and asks me fair and square if I wouldn't like to turn a bob or two doing an honest day's work in a good cause. Well, of course—and you're ahead of me again—I think it's the old carry-on about slipping under the wire after dark and feeding a couple of sawdust sausages to the favourite in tomorrow night's big race, but this low-sized class of a fella says it's nothing like that at all. 'Youse,' he says to me, 'youse can be a fully qualified investigator on the payroll of the

National Council on Alcoholism, if you slip me a unit and keep your mouth shut about who told ya.'

I nearly fell down. 'The National Council on *what*?' says I.

'Alcoholism,' says he. 'The shakes. The old screamin-meemies.'

I was shocked, to the core. 'But that's a private matter,' I told him. 'The Government might as well ask you how many times a day you brush your teeth. Is there no privacy left at all in holy Ireland?'

'Privacy me foot,' says he. 'It's a national emergency. Did you not know there was sixty thousand alcoholics on this side of the Border and another twenty thousand up North?'

'But of course I do,' says I. 'There's four thousand of them here tonight. There's me brother Sylvester and me sister Bridie and her old uncle be marriage—the one that had the bicycle accident a while back—and there's me old butty the Gooser O'Toole—'

'Them days,' he says, 'is over. 'The Irish Goverment wants to change the national image. We're losing millions of man-hours work a day thanks to the old booze and the trade gap's getting bigger all the time and our productivity's gone all away to nothing.'

'Listen,' says I, 'if you want to talk like that go over and do it in Liverpool where they'll understandya. Productivity me big blue eye. Sure that's English talk.'

He gets a bit short. 'Doya want the job or not?' says he. 'I'm a busy man. I'm only getting half a dollar for every new recruit.'

'And who's recruiting you?'

'He doesn't know it yet,' says he, 'but it's me eejit cousin Brendan in the Department of Health.'

'In that case,' says I, 'you're on. That lad's a soft touch even when he's sober. What do I hafta do?'

That's when he gives me this little notebook and the pen thing. 'You just knock on the first door you come to,' says he, 'and ask the fella or the woman who opens it how many gargles they'd have in the day and how it takes them and why they take a sup at all—and everything like that.'

'Is that all?' says I.

'What more wouldya want?' says he.

'But don't we know the answers already,' says I. 'What else would you do in this suffering country, with the rain belting down morning noon and night and Father O'Fiddle on about the Pill and all us young lads waiting until we're seventy, to be on the safe side, to get married and us tormented at the same time wondering if we wouldn't do better to go over to England and lean on a shovel there. Sure it isn't alcoholism we're suffering from. What we're suffering from is—Everything! —Everything! . . . '

Here, Mick, give us another one. I'm a bit short at the moment, but here's a grand little brand new notebook and a sort of a class of a pen to go with it—

Not the Smaller Flaming Space Ship

'Good morning, Mummy darling.'

'Hello, Petekins. You're up very early. Come and give Mummy a little hug. There, that's nice.'

'Mummy?'

'Not too much chatter, darling. Mummy's reading the paper.'

'That's what I've been reading too. I mean, I've been reading Daddy's. It's a horrid one.'

'Please. Petey. My head's splitting. Just sit there quietly or go away and do something else.'

'Why, Mummy?'

'Because, frankly darling. I'm not prepared to discuss the relative merits of two popular newspapers with a child of seven on a Sunday morning.'

'I don't want to discuss the relative merits either, Mummy. I just want to make the definitive statement that Daddy's is a horrid newspaper.'

'You haven't been reading about those disgusting plays—'

'The whiskery ones? The Beards and the Hair and all that? I read about them when they first came out.'

'I suppose there's no means of preventing you. But it's just a passing phase, Peterkin. It's just like suddenly hearing an uncouth person saying a rude word in the street.'

'Like the Lollipop man when I made him trip over his stick?'

'You didn't *push* him, did you?'

'No. I just said, 'Let's get across, then, you silly old—'
'Peter!'
'Let's get back to what I was talking about, then, could we Mummy?'
'All right, then. But do it nicely, darling. Mummy's feeling just a tiny little bit fragile this morning.'
'A contributory factor might have been the new and economically priced burgundy that you and Daddy are trying as an antidote to the precipitous rise in the cost of living.'
'Petey! For Heaven's sake—where did you learn all those awful long words?'
'From Daddy's newspaper. They're not really awful, mummy. Precipitous only means like a sort of precipice—or steep, really. And economically priced is just cheap. And a contrib—'
'I *know* what they mean, darling. I just don't like to hear you using them.'
'Why not, Mummy?'
'Oh—for God's sake—because you're not old enough. It makes me nervous.'
'Because you're only twenty-five, and you feel the hot breath of the younger generation on the back of your neck?'
'Peter. *Please*—'
'Would you like me to say some more about Daddy's horrid newspaper?'
'NO!'
'Daddy's newspaper is talking about the big Christmas squeeze. It's not fair on us children. Does it mean that the Government's going to make all our chimneys so narrow that Father Christmas can't get down them.'
'Peter—you're seven years of age. Please don't give me the Father Christmas bit. Not this morning.'
'I was speaking figuratively, Mummy. The Government's going to be much too busy rebuilding the skyscraper flats to have time to alter Victorian chimneys. But still there's no doubt they're shaping up to put the bite on Christmas.

They don't want another consumer boom, you see, Mummy.'

'Peter. I must insist that you stop talking to me and go right away. This minute.'

'You're putting your head in the sand, Mummy. We want Action This Day—i. e. that you're to get out there tomorrow, Monday, and buy my presents before Roy Jenkins puts on his ten per cent control, with the result that I get the small-sized Flaming Space Ship it's now the same price that the big one used to be—'

'Peter. Shut *up*!

'But, Mummy, we might just as well all shut up, including Daddy's newspaper, because whatever anyone says everyone's going to go as crazy as ever this Christmas and you'll buy me the bigger-sized Flaming Space Ship, which is only ten guineas, plus ten per—'

'Fred! Before I kill him—take this fearful child AWAY!'

Black and White and Beige all Over

The Speaker, to say the very least, provides an unusual appearance, as he mounts between the lions in Trafalgar Square.

His clothing appears to be selected absolutely at random. For instance, the gleaming top hat contrasts strangely with his boiler suit. And just as strange, in contrast to both of them, are the string of beads around his neck and his hairy goatskin jacket. Furthermore, while he carries a tennis racquet, he also wears football boots. The final bizarre touch to this conglomeration is the fact that the left hand side of his face, and his right hand, seem to have been painted black.

The police watch him in silence. There are 10,000 of them on foot and another 2,000 on horseback, while an unidentifiable number sit in Black Marias, armoured cars and coaches with the windows blacked out. At the moment, overfilling the Square as they do, they constitute the only audience, though a thin chanting and a smashing of glass from the direction of Whitehall suggest that various thinking members of what might be called the civilian population are on their way to join the meeting.

The Speaker holds up his tennis racquet, in an appeal for silence, but in fact the gesture is scarcely necessary, as the only sounds—ones of a varied nature—come from the police horses, and are more or less unpreventable.

'Friends,' cries the Speaker, 'men, women and children, fellow union members, chums of the London Stock Exchange, hawks, doves, racialists, anti-racialists, Heathers, Wilsonians,

Thorplettes and Enoch Powellese—greetings! And a specially warm one to you, brother coppers!'

A group of younger policemen, feeling that they are being in some way provoked beyond endurance, start to make a rush for the podium but are trampled down by a sergeant on a white steed, which is receiving supplementary training in preparation for being ridden by the Queen at the Trooping of the Colour next year.

'Thank you, Sergeant,' cries the Speaker. He addresses himself to some of the less severely injured young policemen. 'Why don't you invite him down off that horse,' he suggests, 'to discuss the contribution made by the police to the protection of the right of free speech.'

The Sergeant immediately disappears in a welting of flashing truncheons and helmets swung by their straps. Several of the older policemen upend the white horse and sit on him, resting their feet. It looks like being a long, hard day.

The Speaker seems momentarily upset by the effect—it looks almost like violence—of his intelligent and democratic suggestion, but resumes after a moment.

'As I stand here today,' he cries, 'I speak not only for the workers of this country, but also for the financiers. I come to plead an behalf of the underprivileged white collar worker and also on behalf of the Kenya Asian white collar accountants. I stand four-square behind the millionaire property dealer and the old-age pensioner whose gas has been cut off. I'm for L.B.J. and Ho Chi Min, and I am also for President de Gaulle and his second and third men. I'm entirely in favour of professional tennis players at Wimbledon and will defend to the death the right of amateurs to play on the same courts with them, whilst keeping a different score and playing with the larger ball. I believe that South Africa should be allowed to take part in the Mexican Olympics, provided that all their coloured male athletes are really women, and that Malcolm Muggeridge, under properly observed conditions, should be permitted to take the Pill, but only if this be his own free and unalienable choice!'

The massed police are stunned into a deeper silence than ever. Even the horses make no sound.

'And I believe,' roars the Speaker, 'that Melina Mercouri should be allowed to return to Greece, provided that they make her a Colonel, and that Barbara Castle should be elevated to the Upper House under the title of Lord Castle and that professional footballers should be permitted small arms on the field of play under a strictly regulated licensing system and most of all I believe that all bitching and bawling and protesting and quarrelling should cease, by common consent, for one week, as of today, so that we can think, in place of shouting.'

The massed police divide by instinct. Through them march Tarik Ali and his Action Men. Outraged, they take the Speaker apart and feed him to the loins.

A Good Bog of Public Relations

'GOOD morning. I mean—good afternoon. Well—good evening. This is your Captain speaking.'

The passengers look at one another with some anxiety. They've been doing this in the Departure Lounge since breakfast time, so that they know and loathe the sight of one another's faces, but this is something new. Once every hour they've been told that their flight has been further delayed owing to 'operational difficulties', but at least they were on the ground when it happened. Now they're in the air and by the sound of it their Captain doesn't know if it's Easter or a week ago last Tuesday.

Some of the more experienced among them begin to ease their dentures, and to remove sharp objects from their pockets. The Captain's voice continues:

'I'm most frightfully sorry, ladies and gentlemen, but the fact is that I went to the wrong airport. Ahah-ha-ha.'

No passenger even smiles.

'Everyone seems to be making a bog of it these days. First of all my old chum Captain Seymour oversleeping and taking off forty minutes late for Manchester and now here's me rushing off to Gatwick when I should have been at Heathrow. Still, it's best to be frank. I mean. B.E.A. said Captain Seymour had done a great public relations job by being so honest with his lot so I'm doing the same for you.'

The passengers have sunk their nails into the upholstery of

of their arm-rests and are staring fixedly at the back of the seat in front of them.

'Actually, we do seem to have another little complication. There's no real cause for alarm but I think I'm in the wrong plane.'

Two elderly women start tearing at the emergency exits but are beaten back into their seats by an air hostess, clutching a paper bag in the other hand for her own use.

'It was the rush, you see. Well, it could happen to anyone. By the way, who are you, mate?'

This question is scarcely audible, so that the passengers can only presume that it had been addressed not to them but to their co-pilot. The Captain's voice comes through strongly, and frankly, again:

'There you are, you see. I *am* in the wrong plane. My co-pilot ought to be that great old scout Charlie Horsefall, but this chap says his name's Julian Thesiger. Well, now we are in a muddle. Do you happen to know where we're supposed to be going to, Jule?'

The passengers wait tensely. They can no longer remember themselves.

'You don't know! But you must. *You think you're in the wrong plane too!* It was all that rush? Brother, don't I know it—'

There is the sound of heavy breathing over the inter-com., broken by a sudden cry of, 'Gaw!' The Captain's voice resumes, not quite steadily:

'In the interests of good public relations I must tell you something else. I can't imagine how I got this thing off the ground because I'm a turbo-prop man and now I find it's a jet!' The last word emerges in a near-scream, but the next sentence is a full-throated one. 'And Jule's never driven one either!'

Total panic among the passengers. Life-belts grabbed, twisted on anyhow. Sobbing, prayers—then, suddenly, a miraculous hush. A stocky figure is walking purposefully up the aisle. He wears a distinctive looking short mackintosh and, contrary to the regulations, is smoking a large pipe. He is followed by a small, bald-headed person who, even in this moment of emergency, seems to be taking dictation.

The man with the pipe says, 'Try these for size, Ger. After "I'm a Berliner too"—"The State is also me"—"No Surrender by me either"—and if you can find a suitable translation for "Reich" add on to it, "And again only one Leader".' The silence in the plane is absolute as they pass into the cockpit and the small man closes the door.

For the next hour a familiar voice is heard on the inter-com., giving authoritative instructions to the pilot. The words, 'Meaningful—purposive—technological advances—' are predominant.

The plane makes a bumpy landing. The passengers, hoping for Majorca, find that they're in Bonn. Nonetheless, they are grateful to their *Deus in Machina.*

No one's got where they wanted to get to, but at least they're still alive.

Another great public relations job has been done.

Don't Touch It—It Starts

'Good morning, Skipper.'

'Morning, Sir Basil.'

'Well, this is it. This the great day.'

'Yerp.'

'I must say, we could have done with a spot of sunshine.'

The Skipper feels there is no point in going to the trouble of
agreeing with this observation. The snow, which fell steadily
until the middle of May, was followed by three weeks of
torrential rain and now, as it has been for the last fortnight,
the whole of Southampton Water is enveloped—on Mid-
summer's Day—in dense yellow fog.

'Still, we've got the radar and all that, so we shouldn't
have too much trouble.'

'No, sir.'

'Um. Ahem. Er. Gulp. Engines running?'

'Well, no, sir. Not yet, sir. Actually, I wanted to talk to you
about that—sir.'

Sir Basil involuntarily snatches at part of the bridge equipment, a brass object roughly circular in shape. It comes away
in his hand. He slips it back into place. It falls off. There is a
crash of breaking glass. Roughly, the Skipper kicks the debris
into a corner.

'Sorry about that, Skip.'

'Never mind, sir. What's a compass in a leviathan of the
Atlantic, programmed and computerised to be as modern as
the year after next? Ho ho.'

'We don't want any of that.'

'Beg pardon, sir.'

That's all right, Skip. We're all a little tense. But don't you—um—ahem—gulp—think you could start her up?'

'Well, the way I look at it, sir, there's no great rush. It took Mr. Furcat long enough to get his lot going.'

'Mr. Furcat?'

'The Frog gentleman, sir. The one that got his load into the air about a year late, at an additional cost to the taxpayer of a couple of hundred million quid—'

'M. Turcat, Skipper. And *not* a Frog gentleman. Leading test pilot, if you will, to the brilliant Sud Aviation de France—'

'Whoever he is, he had the right idea. 'Start her up,' they said to him, and he said, "Not on your nelly. Before I start her up," he said to them, "I want a dry or almost dry runway, no tail wind, and clear visibility with a cloud base of at least ten thousand feet. Or otherwise," he said, "napoo." He had the the right idea.'

'At least he got the thing into the air.'

'They talked him into it, sir. That's where he went wrong. What I say is—if you don't touch these enormous things they won't bite. Good old Furcat doesn't know if his sonic boom won't level every skyscraper in New York—after he's melted the polar ice-cap and drowned the whole of the British Isles. Like me, sir. How do I know, if I start her up, that those turbine blades aren't going to go madder than ever and we take off from here, tearing our hawsers out by the roots, and go clean through the Isle of Wight, shaking 'em up in the Royal Yacht Squadron? And if I can't stop her before we get to New York and old Furcat's been there before me and filled the port with the Empire State Building and I finish up in the middle of Connecticut—It's not worth it, sir. Better to leave her alone. Tied up like this, she's a miracle of modern design. Start her up and you've got yourself a right dog's dinner. And there's another thing, sir.'

'What's that?'

'Me runway's all wet, there's no wind at all, I can't see a damn thing, me Uncle Dicky's out there in a rowing boat fishing for dabs and if I run him down his missus will never forgive me. Best leave well alone.'

'You realise that by this attitude you are holding up the onward march of the magnificent British shipbuilding industry?'

'But the Isle of Wight's still there.'

Sir Basil and the Skipper stare out into the fog. Gloomily, they realise that there is something to be said for both points of view but not, perhaps, really enough for either.

Gullible de Trop at St. Trop

IT is round about 7 o'clock in the evening, the time in St. Tropez when everyone saunters up and down the quay, looking at the yachts.

To save space the yachts are moored stern-in on the quay, so that the passers-by can get a good look at the general decor of the various deck-saloons, as well as the quality of the owners and their guests.

Quite a number of these strollers are yacht owners themselves, intent in a gentlemanly way on seeing what kind of mess the others have made of their furnishings this year, or what ghastly kind of friends they've got themselves stuck with.

Once up and down is usually enough, and then everyone retires to their own boat, to sit on the after-deck and to discuss the catastrophes that have overtaken the others. For this ceremony one wears one's smartest summer clothes, and smartest summer manner, in case some of the people taking it in from the quayside are yacht owners themselves.

On many of the after-decks standing-up cocktail parties are going on, with white coated crew members moving among the guests with trays. The English Language, in its more strident aspects, lies like a tangible layer of sound upon gatherings of this kind.

Next door, identifiable by a couple of Ferraris and a Maserati the quay, are a group of languid Italians accompanied by beautiful woman dressed not so much for boating as for the theatre in New York. They lounge about in extremely expensive cane chairs, conveying by their exquisite indolence the fact that they are well accustomed to millionaire yachting and to the heat of the Mediterranean night—unlike the English who have to stand up and yelp about it.

The English often feel like bunging a tomato at them.

Next down the line is an American yacht. Three elderly men sit on the after-deck, smoking cigars. Their wives must be ashore, picking over scarves and costume jewellery in one of the countless boutiques.

If you're English, or Italian, you always feel that cigars before dinner are not quite on and that in any case a cigar aboard a yacht is perhaps just a little too much.

Further down is a French boat, filled with very brown young men and very brown young women. They are drinking very chic whiskeys on the rocks. One feels that somewhere or the other they have been bathing all day in the nude. In an absentminded way they fondle sections of one another as though they were still wearing no clothes. They are probably waiting to go up to Bardot's pad or she is coming down to them.

The other nationalities, apart from the Italian men, try not to watch them too closely. There is a general feeling that what they are doing is not really yachting at all.

Seven o'clock in the evening in St. Tropez, and there we are sitting on the after-deck drinking Pimms, with plenty of fruit and vegetables in our glasses so that the chic-ness of our refreshment will be readily identifiable to the watchers on the quay.

'Did you see what that dreadful old Turk has done this year?'

Maggy, our hostess, has been ashore all afternoon having her hair done by her favourite Jean-Claude, to whom she always goes when she's in St. Trop. He's done her proud, a great mound of pinkish spun sugar, immovable with lacquer in the evening breeze.

'He's actually gone,' she continues, 'and built himself an Elizabethan library in his deck-saloon. Panelling and leather-bound books and all that sort of thing. And, my dears, do you know—he's actually got an open fire as well! With huge logs all blazing away—on an evening like this.'

Richard, her husband, joins in. He spent a lot of time this afternoon on the telephone to his broker in London. The telephone is on the bar in our favourite restaurant. You can see it from the street. I passed by several times and he was still there, speaking urgently into the phone, wearing rather a small yachting cap.

'The really funny thing was that he was wearing a smoking cap—an extraordinary looking plum-coloured arrangement with a tassel hanging down over one ear—'

'And he was smoking, too!' cries Maggy. 'A huge curved pipe like—like Sherlock Holmes.'

'Learning the English way of life,' says Richard, smiling comfortably.

Everyone laughs.

Wolf, the German deck steward, appears. He sidles forward and breathes urgently into Maggy's ear.

Various attempts have been made in the past to call Wolf 'Volf', which is the way he pronounces it himself, but they've been abandoned. There was a lack of consistency about it which led to too many misunderstandings, so that he is now simply called Wolf. He doesn't however, look like a wolf. He looks like a very nervous, pale blonde young German with slight homosexual tendencies, if he ever had the courage to give way to them. Much more of a Volf, in fact, but that cannot be helped.

'Oh, Wolf,' Maggy says petulantly, 'don't be tiresome. The chef will simply have to turn it down. We're not nearly ready yet. And we want some more ice.'

Wolf shivers away. Some of us have got sympathy for him. The chef is a large, dark-complexioned, passionate man who may be either Rumanian or Greek. At all events, no one understands whatever language it is that he speaks, so that he is driven to communicating with his fellows in rudimentary French.

We can hear him at meal-times, because the galley is right below the dining saloon. From time to time he roars, 'Christoos!' or 'Merde!' or 'Putain!' He must be very frightening in that enclosed space.

To prevent Maggy getting on to the servant problem I say to Richard, 'Do you remember the time we went into Tangier and it was pissing rain and you sent Wolf ashore with ten quid to buy some drugs?'

I became aware, a second too late of the presence of the Colonel and his lady. They are visitors from another yacht and while they're not yachting they live in Wiltshire, where the Colonel is Chairman of the local Bench.

I am already regretting my description of the weather conditions in Tangier, and now feel I should take the sting out of the rest of it.

'We only wanted something light, like adulterated keef. None of us had ever taken drugs before and we just thought it might be fun—with the rain, and all that.'

'I see,' the Colonel says, without, however, pardoning the contemplated offence.

'Well,' I rattle on, 'nothing came of it in the end because Wolf came back without any. He said the price had doubled since he'd been in the drug-smuggling business and he didn't want to waste Richard's money.'

Several seconds go by before Richard says, 'Who's for another Pimm?'

After this we settle ourselves again and Richard and the Colonel begin to discuss the possibilities in the Malaga area. They are wary, however, of the unreliable Franco, the Gibraltar situation and the prohibitive charge on property dollars.

Commercialism of this kind seems to me to be out of place

on this lovely yacht, in glamourous St. Trop and on this hot Mediterranean night. Anyway, I'm quite gay on the Pimms.

'What about the time we were in Malaga,' I say to Richard, 'and you and I and Maggy and that curious financial friend of yours set off in a horse-drawn cab to look for a night-club called the Pam-Pam. We jiggled through Malaga, bowing and waving like Royalty to our many fans and admirers, and when we came to the Pam-Pam it was shut. Bolted and barred with the shutters up. We asked the driver what had happened and he said, 'Il Padrone e muerto,' or however it goes in Spanish. And your curious financial friend said, The boss has kicked the bucket? Well, ask him why his sobbing wife can't carry on?, And you in your rough Spanish put the question to the driver who replied by pointing up to Heaven with his whip. We got it. It was Good Friday and by 'Il Padrone' he was having reference to Our Redeemer—'

In that instant I remembered Maggy's warning about the Colonel's lady, about her close relationship to the Duke of Norfolk and her devout, if high, Catholicism, but it's too late.

The Colonel's lady appears to be about to rise to her feet, with the intention of departing, but the yacht lurches so suddenly that she is thrown back into her chair.

The lurch has been accompanied by a loud crash and this is now followed by a passionate cry, 'Jaysus—me foot!'

Everyone runs up towards the bow. I bring up the rear. I don't like the sound of that accent, and with reason, as it turns out.

With the propeller still churning away in reverse a dingy little ketch is grating away at the immaculate white side of Richard's yacht. The ketch might also have been white, but many many years ago. A line of indescribable laundry hangs from a piece of string between the mast and the forestay. Amidships is a young man with flaming ginger hair, wearing a pair of filthy flannel trousers and nothing else. His bare foot does indeed appear to be wedged between the yacht and the smaller boat.

'John-Joe,' he bawls again, 'me foot. It's cot. Will yez knock off the —in' ingin for Jaysus sake—!'

The engine stops. The smaller boat swings clear. The young man, cursing foully, tenderly nurses his foot with one hand. He looks up and sees me retreating into the wheelhouse.

'Holy Mother!' he cries. 'It's Paddy. Paddy Campbell from Dublin.'

Wild with excitement, his injured foot forgotten, he shouts down into the ketch's cabin, 'John-Joe—Norah—Bridie—come up owa that! It's Paddy Campbell. The hard man himself. Doin' the lord on this—in' big t'ing here!'

His companions appear. To judge by their appearance they've been at sea for a long, long time, and it's been an unusually tough trip. Jean-Claude, faced with the matted locks of Norah and Bridie, would have had the vapours.

But all of them are delighted to see me. 'Holy Mother' and 'Suffering duck' ring out in the warm night.

John-Joe remembers that first things come first. He has a dark grey hand on the gleaming teak rail of Richard's yacht.

'Hey, Paddy,' he says urgently, 'is there air of a chanst of a botla shtout? Jaysus, we're—in' pairched.'

The Colonel and his lady have withdrawn to the safety of the stern. Maggy's in the wheelhouse, pretending it isn't happening. Richard says to me sternly, 'I shall be glad if you will tell your friends to pull away. The harbour regulations do not permit double-berthing.'

As it turns out a moment later there is no reason for him to be so toffee-nosed because there is a sudden roar of rage from down below, followed by a scream and Wolf erupts on to the deck as if propelled from a gun. His pale blue eyes are rolling in terror. He grabs Richard by the arm of his immaculate blue blazer and sobs out, 'Cook—epileptic attack—hess knife—Herr Gott—Herr Gott—!'

The evening peace of St. Trop—tenuous at the best of times —has been shattered, at any rate in our section, for at least another twenty-four hours.

Blunged on Vidka

WE met a charming Russian couple the other evening, and a quadrilateral fusion of interests broke out. None of this 'love him, hate her' stuff. They were really delightful. Intelligent, imaginative, amusing—all the desirable qualities.

He was dark and she was fair, and their English was only slightly accented. They asked us to dine with them. 'We will eat blinis,' he said. 'We will have a true Russian dinner,' she said. 'It will be very nice.'

We presumed that they must have something to do with the Russian Embassy, and began to look forward to blinis carpeted with the best caviar.

When we arrived at their flat nothing happened, however, for quite some time. As often happens in these cases, the blazing new friendship was not quite as incandescent as it had been before. The conversation was somewhat constrained. Nor was there any trace of refreshment. I had begun, in fact, to think we'd stepped into a nest of teetotallers, when he suddenly seemed to remember his duties as host.

'Now,' he said, 'we drink vodka.' From a cabinet containing only a limitless number of bottles of vodka he selected one and placed it with four very small glasses on a tray. He filled our glasses and said, 'Straight down, in the Russian way.' We did so and he immediately refilled them. I allowed a couple of minutes to pass before having another Russian go. He refilled my glass. I emptied it as we rose to go into dinner and he re-filled it. I emptied it and he refilled it. I emptied it again and he refilled it and then, following my wife's thoughtful example, took the full glass with me into dinner.

On the table was a large silver chafing dish full of large pancakes, or blinis. There was a big silver jug of melted butter and another of sour cream. There was also a long tray of things like smoked mackerel, anchovies, pickled herring and so on. No caviar.

Inadvertently, while waiting to begin, I emptied my glass. It was refilled immediately. Our host then explained the pro-cedure.

'Put a blin on your plate,' he began, and then she said, 'One blin, two blinis,'

'Put a blin on your plate, pour some butter and cream over it and then fold in piece of mackerel. Then you have more vodka. It helps with the greasy butter.'

The thought of greasy butter caused me to take advance precautions. My glass was refilled. I noticed that my face was getting a little stiff and that my head seemed to be swelling.

We started on the blinis. Very nourishing. Almost, perhaps, a thought too rich. I had to pause half way through my third blin to emulsify with vodka and my glass was refilled.

By now I was really getting into the Russian way of life. I had another vodka *before* my fourth blin and said, 'Surely we ought to say something Russian when we drink. Something like, Krasnayia slovolovitch!, or whatever it is. What it is?'

'I don't know,' our host said. 'I don't speak Russian. I left when I was three. Krasnayia slovolovitch!' He refilled our glasses.

'I don't speak Russian either,' his wife said. 'I was born in Lille. Krasnayia slovolovitch!' Our glasses were refilled.

This time, when helping myself to another blin, my glass was momentarily empty, so that I had the misfortune to fill it with melted butter.

The error threw me into a state of despondency. Almost crying, I said, 'I'm cut off from solovolovitching. I've blindered.'

'You've blunged,' our host said with some severity. 'The verb is irregular. I blinder, thou blinderest, he blinders, we blanger but you blunge. Krasnayia slovolovitch!'

Not to be left out of the slovolovitching I poured the butter back into the sour cream jug, washed out my glass with vodka and poured it over my blin. The I put another blin on top of it, instead of a slice of pickled herring and almost broke down. 'I can't slovolovitch any more,' I cried. 'My syncronisation's all gone to blunge.'

The next thing I knew I had a cup of tea in front of me with a slice of lemon in it, and beside it a plate of jam. My hostess put a spoonful of jam into her tea, but I got mixed up and poured the tea into the jam.

Next morning my wife told me I'd invited our new friends to a real Irish dinner. 'Kedgeree Kathleen Mavourneen you called it,' she said. 'Do you know the recipe?'

By way of answer I turned my face, the size, shape and colour of a pickled gherkin, to the wall.

Mr. Egg and the Long American Legs

Madame came out into the garden round about 7.30 in the morning and said, 'Mr. Egg has just been on the phone. He sounds agitated. He's coming round at once.'

It seemed unlike Mr. Egg—a cool and elegant man of about fifty. He lives alone in a charming old farmhouse back in the mountains, and has never been agitated about anything, as long as we've known him. He even likes being called Mr. Egg, or M'sieu Oeuf, the closest his French neighbours care to get to his rightful christian name of Hugh. We are very fond of Mr. Egg. He's extremely good on sub-tropical gardening and the preparation of new plaster for painting and the making of fish soup and many other social graces of that kind.

When he arrived, half an hour later, he brought with him a bottle of chilled champagne and a jug of fresh orange juice—a combination known in these parts as a mimosa. As he poured he said, 'A little early, but then the situation is tense.'

We toasted one another silently. Then Mr. Egg said, 'I've just been to New York. There I met a lady called Amanda. Divorced, thirty-nine, two first-class, long American legs, beautiful Park Avenue penthouse, well read, extremely chic and absolutely childless. We saw a lot of one another.'

That's another thing we like about Mr. Egg. When serving he does it quickly, neatly and in just the right amount.

'Four days ago,' Mr. Egg went on, 'she phoned me from London, where she had arrived unexpectedly. I asked her to come down here and stay with me.'

After a moment Madame said, 'Why not?' It was a way of telling Mr. Egg that that part of the business, at least, was quite rational.

'The following morning,' said Mr. Egg, 'at 11.35, when she got up, I found she couldn't drive a car. She could not conduct,' he said, choosing the words with care, 'an automobile. A chic, long-legged American woman, well read, childless, who could not drive a car, never had and did not propose to begin now. 'I found out,' he said, 'she couldn't drive a car because she wanted to have her hair done in Nice, at 11.35, at an address given to her by a friend in New York.'

Mr. Egg diluted this vision of hell with a little more mimosa. 'I explained,' he went on, 'that the establishment would be closed from midday until 3 p.m., that it would be fully booked out because the new French travel allowance has deposited the whole of the haut monde of Paris on the Côte d'Azur and that in any case I couldn't possibly drive her there because I had been waiting for three weeks for the arrival of the plumber, Jean-Claude, and that if I left the house even for ten minutes he would certainly appear and then go away again for ever.'

Mr. Egg paused, with bowed head and closed eyes. 'I offered,' he said, 'to ring the hairdressing establishment and, if an appointment were available, to get her a taxi that would leave her at the door. She refused this offer,' he said, 'on the grounds that she didn't want to be alone in a taxi with a French taxi driver for the reason that she could not speak French, never had been able to—'

'And did not propose to begin now,' the three of us said in unison.

'Instead,' Mr. Egg said, 'she wanted a dry Martini made with gin. Of course I had no gin so she had the last of my last bottle of whisky instead. At lunch she couldn't drink the economical wine I've been drinking for years and had three brandy and Perrier in its place. After lunch she wanted to go down to Cannes, to the beach. I explained again about the congestion of the coast in August and that in my case, during the heat of the day, many of us like to lie down with the shutters closed. She didn't want to do that either.'

Mr. Egg sighed. 'Before dinner,' he said, 'we walked down to the Bar Napoléon in the village, looking for a bottle of gin. They didn't have one but Jean-Claude was there with his brother, so we had a pastis or two but she just sat there, tapping her fingers on her knee. She said the place was dirty. As of course,' Mr. Egg said generously, 'it is, but it's the only one we have. After dinner,' he said gravely, 'she wanted to go night-clubbing. It was nine-thirty p.m. I could hardly keep my eyes open. She sat up for hours after I'd gone to bed, trying to get Radio Luxembourg on my little transistor, but apparently there was only a lot of yelling in Italian.'

He thought for a moment. 'She's asleep now,' he said, 'but what am I going to do when she gets up?'

'Nothing,' Madame said, coming to one of her well-known, firm decisions.

That's what he must have done, because he drove past here yesterday alone save for a large sack of fertiliser on the seat beside him.

He tootled the horn and gave us the thumbs-up sign. It had all come right in the end.

Desperately, All Morning

THE voice cried piercingly over the garden wall, 'Darlings—been trying *desparately* to phone you all morning!'

It was tossed over the wall in the way that a hat, in the im-memorial bar-room tradition, is thrown into a saloon, to see if it's going to be thrown out again. A kind of sighting shot, a testing of the temperature of the water.

Madame and I said nothing, shelling beans quietly in the sunshine. But we exchanged the kind of look that passes, perhaps, between two hard-bitten Commandos, snug in their foxhole, whose trained ears detect the advance of trouble.

And trouble, of course, it must be, because the voice was the voice of none other than the toast of the Coast—the (unfortu-nately) irrepressible Dinky Scrimgeour-Scrivener, so far off her normal Antibes-Cannes beat, up here in the mountains, that she must have some very special request to make of us, one that would cost (us) quite a lot of time and probably money to fulfill.

A moment later it was plain that the matter was urgent, because suddenly Dinky was right in amongst us, without even the small formality of touching the goat bell at the gate. 'Darling!' she cried again. 'Fearfully sorry banging in like this, but I was so near I just had to.'

At this the miniature poodle in Dinky's arms went off like a minature siren—a thin scream that pressed our eardrums together in the middle. Dinky paid no attention to it whatever. Through the scream she gave us her wistful, little-girl smile, the one that asks to be forgiven for naughty presumption—a considerable technical achievement for a lady in the middle, restless fifties.

'That's all right,' Madame said in her level kind of voice. 'It's nice to see you.'

Both of us were trying to estimate, with a speed of computers, what it was that Dinky wanted; or what it was that we had of which she required the use. And, of course, to head her off before she got us jammed in a cleft of non-existent moral obligation.

In a split, silent second we came to the following conclusions, though not necessarily in the same order.

Dinky probably didn't want to remove Madame T., the lady who does for us, nor Marcel, our handyman. To get them to her villa on the coast, and home again, would have occupied the services of one of her three cars at a time when it might be needed for something more important.

Dinky probably didn't want to remove our builder, and his three remaining workmen. Her villa is constantly being repaired and enlarged by a resident team of masons, plumbers and electricians and it seemed improbable, even if she wanted a few quick shelves put up in the servants' hall, that she would go to the trouble of stealing our lot.

She couldn't want to borrow our car, our cat, my pick-axe or Madame's silk suit, and yet here she was—all the way from the coast and a little on edge. In fact, she was so on edge that she suddenly silenced the poodle with a rap on its lug and came to the point, albeit with a number of detours.

'You see, darlings,' Dinky said, 'ever since we got back from Capri, Alexis has been terribly difficult. You know?'

We didn't. We'd never heard of Alexis and could only guess that he must be new.

'So the darling boy,' said Dinky, 'saw this advertisement in some English newspaper for some marvellous new kind of lawnmower. It floats in the air or something. And so he wanted one immediately.'

Alexis—a new and rather special gardener?

'But the trouble is,' said Dinky, 'they cost twice as much here as they do in London.' Then she played her ace, pretty quickly. 'So I thought,' said Dinky rapidly, 'that as your dear little daughter is driving down here fairly soon with her little friend she could possibly just slip one of these lawn-mowers into the back of the car and of course it's going to be difficult for you with so many girls but Alexis has masses of the most divine boy friends and of course I'd settle up with you later.'

Madame settled up instantly. Back of daughter's car already bursting with contraband, etc., etc., leading to extreme shortage of cash, etc., and in any case divine boy friends perhaps not quite ideal for man-hungry English girls . . .

Dinky left before Madame had finished. We went back to shelling beans.

'It's always so sad,' Madame said presently, 'when they think you can't see them coming.'

Shorter and Filthier in English

The last thing one would want to do, of course, is to stir up inter-racial feeling or discrimination or anything like that, but one has to allow that one of the major problems created by settling abroad is that of integration with one's fellow nationals, particularly those who saw Labour efficiency coming at home and got out from under it some years ago.

These veterans of the emigré life on the Riviera are settled in their ways, and have such mountains of dirt on the other ones that it's difficult for the newcomer to find his way through the social scene without giving offence, or appearing to be a fool.

A very light commendation, perhaps, of the near-prettiness of the wife of Major 'Hot' Chutney leads to faces being closed all round the room and in no time at all we're loudly discussing, once again, the possible devaluation of the franc, and what we can do to hold up our remittances from the City of London until that happy day dawns. It is later revealed, in a confidential aside on the telephone, that Babs—Mrs Chutney, to you—once borrowed ten frances, during the course of a Bridge game, from Lady Glue, and not only failed to pay it back but also omitted to write to thank Sibby—to you, Lady Glue—for her kindness in arranging the Bridge party in the first place.

It takes weeks of being exceptionally thoughtful and hospitable to everyone to work out that kind of misdemeanour, and even then one never gets entirely clear. Another false, step, like splashing Gerry's glasses in his very own pool, puts the ten francs and the missing letter right back on the charge sheet

again, for detailed examination by the jury, and certain punish-
ment to follow.

There is another difficulty for the newcomer in this social
swirl. Almost everyone has names like Sibby and Gerry and
Babs, and even Lulu, Timmy and Mab. There is a feeling of
joie de vivre about them—a whiff of yachts and Ferraris and
that rather special kind of naughtiness that comes over the
English upper crust in a warmer climate. So that when someone whom we have actually invited to have a drink rings up
to ask if Timmy and Lulu and Mab can come too there is a
tendency to feel that such gay-sounding companions will be
more than welcome.

Until, that is, they arrive. Lulu and Mab have extreme
difficulty in getting out of the car and Timmy can't make it all.
The recent wet weather has played Old Harry with his gout,
rheumatism and arthritis and Lulu and Mab are very little
better.

It would certainly be unforgivable to blame them, indivi-
dually, for being over seventy years of age, but one cannot
help feeling at the same time that their Christian names have
preserved a certain disproportionate fluffiness, out of tune
with their actual condition.

Even in this part of the world, where the village of Opio
is known as 'English Hill', there are some French people in
residence, however, and it is a pleasure to hear them speaking
in their own language—particularly after yet another session
with Gerry, Sibby and Babs on the subject of the franc. We
met a charming couple the other day in the village and asked
them to drop in for a drink. Both from Paris, speaking impecc-
able French and enjoying the summer retreat which they've
recently bought in the neighbourhood.

Madame took on M'sieu, her French nearly as good as his
own, while I—not quite as silver-tongued—chatted up his
wife, a girl of unusual delicacy, both in appearance and
speech.

After a couple of whiskys, however—the most chic of all drinks in France—she became almost sportive, emboldening me to abandon the subject of the coming elections and to advance into the slightly coarser colloquialisms of the French language.

I pointed out to her that it was difficult for me to build up a sufficient store of them, in conversation with the local inhabitants, as they employed an argot on these occasions which I found not only incomprehensible but also impossible to reproduce. 'On the other hand,' I said, 'your French is so wonderfully clear I'm sure you can help me a great deal.'

Gaily, she asked me what I wanted to know. Over the next few minutes I discovered a number of judiciously risqué things to say to other motorists, venal taxi-drivers and so on, and then suddenly I remembered a phrase taught to me by my stepson, one which he'd learnt early on during his year at Nice University.

I reproduced it for her. 'It's funny,' I said, 'it's so much shorter in English.'

The lady closed her eyes. All at once, she was very pale.

'I do not know what you say,' she said, speaking English for the first time herself. She and her husband left almost immediately.

Society-wise, I should think we're back now in the arms of Timmy, Babs and Sib.

Dinky, Disappearing

WE were engaged upon the gentle pleasure of strolling through the new yacht marina in Cannes, counting the red ensigns drooping in the heat and speculating upon the present activities of their owners.

'That lad there,' I said, indicating a boat that looked like five or six very large American automobiles all put together, 'at this very moment is getting his feet under the lunch table in the Savoy Grill—probably raining very heavily outside—and wondering in quiet desperation how to dissuade his guest from cancelling his order for ten thousand nuts and bolts. Because if the order is cancelled the crew of eight won't get their wages and they'll leave in a body to join another yacht where three's a steadier increment for continuing to do nothing.'

Madame said sympathetically, 'We can only wish him good nutting and bolting,' and we moved on down the quayside, glad to be ashore, only to be hailed a moment later by a female voice that called, 'Ahoy, there—yoo-hoo!'

Adjusting our eyes to the glare we saw that it was none other than the Honourable Mrs. Reginald Scrimgeour-Scrivener, toast of the coast and 'Dinky' to her friends.

In a cloud of ash-pink hair, very tight trousers and a beautiful silk shirt she stood at the stern of a boat no larger than two Cadillacs, attended by three very bronzed males wearing yachting caps so carefully crushed to a casual shape that they had to be professional seamen.

'Yoo-hoo!' cried Dinky. 'Do come aboard.'

There seemed to be no way of avoiding it. Madame, who suffers from vertigo while stepping off a low curb, made her way rigidly across the narrow gang-plank, unassisted by the professionals. In the yachting business professionals, while just able to tolerate their owners, can never see why the owner's guests shouldn't be chopped up for dog food, and they show it, too.

'Well—lucky you!' cried Dinky. 'We're just off to the islands. You must come. Blissy bathing and drinkies—'

She was interrupted by the arrival, at 70 miles an hour, of a young man with very crinkly black hair at the wheel of a scarlet Alfa Romeo convertible. He stopped, with a screech, an inch from the sea and blew his horn peremptorily.

Dinky waved to him gaily. 'Shan't be a sec. darlings,' she said to us, and left. The back wheels of the Alfa smoked as they disappeared.

The three professionals looked at us for a while through big dark glasses, and then they went, too.

We sat on the after-deck in the sunshine for perhaps fifteen minutes, wondering what to do. 'We could leave a note saying you'd been fearfully seasick,' Madame suggested, 'if we had anything to write on, or with.'

'You could be seasick too,' I said, feeling injured. 'It'd make it more convincing.'

Ten minutes after that there was a burst of laughter from a boat further down the line. Rather a large cocktail party had suddenly materialised aboard it. In the middle of it was Dinky, animated as a catherine wheel. We were trapped. The cocktail party was between us and the entrance to the marina. Any attempt to get past it must be greeted with more yoo-hooing, and an even more disagreeable involvement.

'What about a blissy drinkie?' I suggested. 'That cabinet thing there must be bursting with it.'

'I tried it,' Madame said. 'It's locked.'

We continued to sit on the after-deck, in the sunshine. There was nothing to read. Not even a piece of string, to tie into knots.

'I can't believe this has happened,' I said, after a while. 'It's literally incredible. I mean, we used to be so happy before all this broke out.'

'Just think about it steadily for a while,' Madame said. 'You'll get used to it. I have.'

Ten minutes late we heard the thunder of powerful engines. It was the cocktail party yacht, putting to sea, with the cocktail party still going on. There was, however, no sign of Dinky.

She, it turned out, was having blissie drinkies with the furry young Alfa Romeo man on the terrace of the Moby Dick bar, at the marina entrance. She waved to us, rather briefly, as we went by.

'Will I give her a yoo-hoo?' I asked.

'Don't touch it,' Madame said forcefully. 'Just come on.'

Two and a Half of His?

A man whom we know vaguely brought another man to see us the other day whom we didn't know from Adam.

The stranger wore an excellently convoluted Panama hat. It was mature and as individual to him as the bullet-holed, sweat-stained, dusty old stetson of the television trail boss. A small work of art, on its own.

The rest of him was very good, too. Sun-bleached alpaca jacket, with a pair of secaturs in the breast pocket and a few skeins of raffia trailing from one of the side ones. Baggy old corduroys, even in the heat of the southern summer, and a stout pair of brown walking shoes. And the whole thing set off by a red and yellow silk square round the throat. I. Zingari? At any rate he looked like a clean, spare, lithe, gentlemanly old cricketer who had turned, with dashing Corinthian brilliance, to rather expert gardening in his later years.

It turned out that he was justified in the amount of time he must have spent upon perfecting his appearance, because this was exactly what he was—Fairlie Hardy, from Monte Carlo, an absolute genius with flowers.

As they came in through the gate the friend introduced him in a voice so hushed with reverence that he might have been handling Fairlie Hardy on a professional P.R.O. basis. 'You know,' he whispered to me, after the introduction, 'the Rothschild place on Cap Ferrat.'

I said, 'What's he doing here, then?'

'Nothing at all,' the friend said cheerfully. 'We were just

passing by and he said he'd love to have a look at your garden.'
The friend went so far as to give me a little touch of his elbow.
The right eyelid almost closed. 'He always reads your jottings
in the Observer,' he said.

I was about to deal with this misconception, demolishing
the revolting 'jottings' on the way, when I saw that Fairlie
Hardy, the genius with flowers, was at it. He stood in the middle
of our small, walled garden, in which geraniums grow like
weeds, and his left hand, with the fingers sensitively and delicately curved, was raised to about shoulder level, the finger-
tips almost visibly tasting the air.

After a moment, Fairlie Hardy was able to analyse his
emotions with sufficient accuracy to give them expression.

'Persuasive,' he said, giving it the weight due to the only
possible word.

Madame has a special look that she gives people on such
occasions as this. It's quite narrow, perhaps nearly steely. She
doesn't speak, but the head is placed a little on one side. Her
attitude says, very clearly, indeed. 'Kindly wash out your
mouth and then, omitting the rubbish, begin again.'

Fairlie Hardy got it straight between *his* eyes. He cleared
his throat. 'I mean,' he said, 'it speaks to me. It tells me—I
like it well.'

'That's good,' Madame said, closing the subject like the door
of a safe. But Fairlie was courageous. 'Perhaps, though,' he
said, 'just a hint too great a profusion of the Pelargonium
Peltatum. Might I suggest Trachylospernium Jasminoides?'

'We had some of that,' Madame said, 'but it died.'

They fought it out steadily from then on, toe to toe. Fairlie
said the new lawn was full of Mullumbimby Couch Grass.
Madame said she liked the colour. Fairlie said that one of our
mimosa trees—he called it 'Acacia podalyriaefolia'—was going
to get much too big for the place it was in and Madame told
him that before that could happen she would 'lop it until it
squeaked'.

During Round Nine the friend said to me, 'I told the old goat it wasn't going to work.' He was resigned, rather than bitter.

'You mean,' I said, 'Fairlie Hardy thought he was going to remake the entire garden, seeing that we're new here and he hoped we wouldn't know any better.'

'He wouldn't do anything himself,' the friend said, defending Fairlie's reputation, 'but the nursery in Cannes would give him ten per cent of their nett for the introduction.'

'And what's yours?'

'Two and a half of his,' said Fairlie's friend.

They left, after drinking a whole litre of wine each.

We gave our Pelargonium Peltatum a good watering that evening. It made the whole garden look fresher.

Gullible from Pen to Philley

SOMEONE knocked politely on the cell door. After a moment, there was the jangling of heavy keys. The door opened and the Superintendent came in. Behind him, with the keys, was the Sergeant.

'Good morning,' the Superintendent said. He looked at the breakfast tray on the bed. 'If you've finished,' he said, 'it will be all right for you to go home. I'm just off myself, as a matter of fact. We've had a busy night.'

He'd shaved, but he looked very tired.

There was a movement in the narrow passage outside the cell. 'Oh, by the way,' the Superintendent said, 'I'd like you to meet my colleague. He'll be—looking after you from now on.'

A younger and leaner Superintendent came in.

I said, 'How d'you do' We shook hands.

The other Superintendent said, 'You will be in court this morning, won't you? Ten o'clock. The West London Magistrate.'

I said, 'I've already explained about that. It's going to be very difficult.'

'I strongly advise you to attend—sir,' the Superintendent said. He didn't want to go into it again. He put out his hand. 'Goodbye, sir.' He went away, followed by the Sergeant.

The new Superintendent and I stood irresolutely in the passage.

To say something, I nodded towards a hand-written notice

stuck on the door of the cell opposite. I'd looked at it several times during the night through the spyhole in the door of my own cell. The notice read: 'The Ice Box.'

'That's not your torture chamber, is it?' I said, trying to smile. The new Super managed a laugh. 'No, no,' he said. 'We've got a lot of stolen refrigerators in there. One of the lads just stuck up the notice for a joke.'

In the middle of my other troubles it was a kind of relief.

He wanted to get on with his day's work. 'Well, then,' he said, 'I'll see you later. Your car's in the garage. If you'd like to follow me—'

A couple of minutes later I was driving home. Five hours before I'd also been driving home, when the young policeman stepped out of the shadows and put up his hand. Later he described it as 'just a routine check'. It was a routine check that landed me in a cell—'over-tired in charge of a motor vehicle'.

Even now, ten years later, I still call it 'over-tired in charge'. The reality is too shameful.

I drove home. The car was exactly the same, but my own circumstances had changed out of all recognition.

Five hours before I'd been looking forward to flying to America for the first time. A week in Philadephia at the expense of Pan-American Airways, which had already flown me so many thousands of miles. But the plane was due to leave London Airport at 10.30 that morning, and if the Law had its way I'd still be in the West London Magistrates Court at that time.

In many ways I'd had as busy a night as the gentlemanly Superintendent. I was incapable of applying my mind to the problem.

There'd been an old and greasy copy of *Woman's Mirror* in the cell. I'd read every word of it and then slept for about twenty minutes, to wake and to begin pacing up and down not thinking about the locked door. Now, I couldn't think about Philadelphia either. There was about an hour and three-quarters before the plane took off, either with or without me. I didn't know which it was going to be.

The moment I got home I rang Madame, the most sturdy person in an emergency. I woke her up, but she grasped the situation immediately. She even had time to say, 'What bad luck. But don't worry. I'll shop around for a legal eagle and call you back.'

I thanked her hysterically. At that moment my own solicitor was probably waking up and yawning in some outer suburb, but both his suburb and his telephone number were unknown. I though I'd once heard him talking about Sevenoaks. Or was it Heywards Heath? It was nearly 9 o'clock. There was no time to track him down.

Still torn between Philadelphia and the Court I was packing when Madame rang me back. 'You're off to Philadelphia,' she said cheerfully. 'I've just caught a barrister in his bath. He says you've only got to write a letter to the Magistrate explaining the position, and tell him you'll be back in four days time, or whenever it is, and it'll be all right.'

'He's sure?'

'He says he's sure.'

'Thank you more than I can say.'

'I won't keep you. You must be busy. Have a good trip.'

I wrote the letter very quickly but the typing was awful so I wrote it again, more carefully, and signed it as legibly as I could. I took a taxi to the court—it seemed easier than the car. There was a policeman outside. I·pressed the letter into his hand, without speaking, and jumped back into the taxi again, in terror of being seen by the Superintendent, thinking he was certain to arrest me. The taxi left me back at the flat. I was waiting on the pavement with my suitcase when the car arrived from the airline at 9.30 a.m.

Someone else was in it. Another journalist, a fat jolly man whom I'd seen around several times before. We exchanged greetings, and drive off in silence. I know he was looking at me with a certain curiosity, but I didn't meet his eye.

Suddenly, I realised we were driving up North End Road and at any moment would be passing with in a couple of yards of the court I'd just left. And, almost certainly, just at the moment when the Superintendent arrived.

I crouched right back into the corner, ducked my head and covered my face with my hand.

After a moment the other man said, 'Not feeling too good? Rough night.'

'The worst,' I said, without looking up. I wondered, in the days and nights to come in Philadelphia, if I'd be able not to tell him what had happened. I didn't want to tell anyone and then, in the next second, I wished to God I'd gone to the court and got it over. It was senseless to go tearing off to Philadelphia. Then I had another, altogether terrible thought.

The Superintendent only had to ring the airport police. I'd be arrested at the immigration desk and probably be given a prison sentence. Three months. I thought of the five hours I'd spent in the cell, the tiled walls, the bright light, the thin blanket on the hard bunk . . .

I must have groaned or made some sound of despair because the other man said, 'Not much further now.' Then he added sympathetically, 'They'll probably give us a nip before we take off.'

I thought of the nips that had got me into this situation, and didn't know if I wanted one or not.

We were given one in the airline's V.I.P. lounge. Several of their representatives were there to see us off. I could scarcely talk to any of them, still shaking from the business of going through immigration. A number of hard-eyed men had been standing around, looking very like detectives. The immigration officer had given me a glance so penetrating that when he handed back my passport I thought he was signalling to the plain-clothes men. I stood in front of the desk, holding my passport, waiting for the hand to fall on my collar, until the immigration officer said, 'That's all, sir—move along, please.'

Evidently, I was going to be allowed to go to Philadelphia, whatever they proposed to do to me when I got back. For a split second I wondered what the rest of my life would be like if I didn't go back at all, if I borrowed some money and went to rich friends in New York. I could get a black-market labour permit—

For that split second the leaden weight of sick misery lifted. I was planning to do something of my own volition, something that hadn't happened since the young policeman stepped out of the shadows and put up his hand. It was gone as quickly as it had come. I knew what was going to happen. I was going to trail round Philadelphia for four days, with the weight getting heavier and heavier, and then I'd return to London where I'd be fined something like £50, and I'd lose my driving licence for a year. If nothing worse happened because I hadn't attended the court . . .

I read a number of English newspapers, as we flew across the Atlantic. In all of them was the story of a well-known actor who'd been arrested in exactly the same circumstances as myself. I'd forgotten about the newspapers. Perhaps because I worked for one the others wouldn't use it. I couldn't even think about it. It was just another thing to be faced when I got back.

When we landed in Philadelphia the temperature must have been in the nineties. Standing in line for the Customs I knew I was going to faint. My knees were going to give way and I was going to fall on my face on the concrete. I'd finish up in some unimaginable hospital and be shipped straight back to London from there, leaving me with nothing to write about Philadelphia and in the airline's debt for the rest of my life.

A policeman in shirt sleeves with black glasses stood in front of me. The butt of an immense gun stood out from the holster strapped to his side. 'March, buddy,' he said. 'We ain't got all day.'

Before I had time to think that he was arresting me I saw what he meant. I was holding up the queue. I shuffled forward, filling up the gap. In comparison with the courteous Superintendent this policeman—this cop—was very frightening indeed. America was obviously not a place to be arrested in.

After Customs the fat, jolly journalist and I stood for quite a time in a huge cavernous hall full of milling people—the men in shirt sleeves and even the old women in light summer dresses. Again I thought of sliding away, of losing myself among them, of never going back to London, but I knew it wouldn't work. America, seen for the first time, was more foreign than any country in Europe, and I looked utterly foreign in it. There was no chance of hiding here.

We were claimed in the end by three genial young men from the airline office. In a car as big as a boat we drove into the middle of Philadelphia. A surprisingly large number of the shop windows were boarded up, and looked as if they'd been like that for a long time. One of the young men said, 'The city centre's dying. Everybody's moving out to the suburbs.'

Under different circumstances I would have been interested, and asked him questions. Now, like the big cop with the dark glasses, it just seemed further evidence of the ruthlessness of America. I couldn't think of anything to say.

'Still,' the young man said, 'we've got some good eating houses left.' He seemed to share a joke with his friends. 'I guess you'll like this one,' he said.

For a moment, in blazing sunshine, I thought we were going to have dinner. Then I realised it must be lunch. We'd left London at 10.30 a.m. But it was probably only about midday in Philadelphia.

In terror, I began to work backwards. I hadn't been in bed, apart from the twenty minutes on the hard bunk in the cell, since the night before last. And now we were going to have lunch and then be shown all over the city. Later on there would probably be a cocktail party, and a dinner party and then a nightclub. If I began drinking again I was certainly going to fall down as much from exhaustion as anything else—and finish up for the second time in a cell.

For lunch I resolved to eat as much as I could, and drink as little as possible.

The restaurant was decorated, successfully, to look like an English chop-house. The waiters wore red coats and knee-breeches—and the manager came from Dublin.

This was the treat that the young men from the airline office had prepared for me, because the manager remembered me well from the old days, and there was a bottle of Jameson on the table to welcome me.

I remembered him, faintly, too. 'Slainte,' we said to one another, and had the first one. We reminisced for a while. After two more Jameson, he said, 'Would you ever come into the back office a minute? The wife's in there and me mother and she'd love to meet you again?'

The proceedings in the back office took nearly an hour. When we left the restaurant at about four o'clock I didn't immediately recognise the fixed grin and the glittering, glassy eye that met me in a mirror.

We called on the City Hall, where I met Police Chief Petersen. 'This guy,' someone said proudly, 'can spot a crime a half hour before it's committed.' I shook him confidently by the hand. He punched my bicep in a playful way and I steadied myself against a table.

We examined the Food Produce Centre and the Schuyikill River and the Liberty Bell. We had dinner somewhere or another and I don't know what time it was when I said to the young man, 'I'm sorry, I've absolutely got to go to bed. You see, I spent last night in a cell in London and I didn't get a lot of sleep.'

He took it very well, or, perhaps he was thinking more of

his own commitments. Outstanding among them was the necessity to get the fat, jolly journalist and myself to a café from where someone called Frank Ford ran an all-night radio show. When we got there, to be interviewed by Mr. Ford, the young man hoped we would be able to mention the name of the airline as favourably and as often as possible.

'Just do this one thing for me,' he said. 'Then you can hit the sack.'

The programme must have been on for some time because the café was jammed with people, many of them in a state of high excitement. At a table on a dais at the end of the room sat a man with silvery-gray hair and a humourous, intelligent face. He had a telephone receiver to his ear, but the voice of the woman he was talking to was broadcast by a loudspeaker on the wall behind him. It seemed to be a general discussion, because the café audience joined in. The subject was public transport in Philly, and no one thought very much of it. Once again America seemed to be raw and violent and ruthless. I wondered, half-consciously, what they would think of us.

The fat and jolly journalist was very good. He'd read a lot about Philadelphia and got several rounds of applause for his enlightened opinions. I was almost asleep when Frank Ford nudged my elbow. I opened my eyes. He was holding a piece of paper. Written on it was, 'Campbell fresh from pen. in London.'

The young man must have passed it up to Ford in the hope of rousing me into taking some part in the show.

Ford said, 'Mr. Campbell—I'm informed you've arrived here straight from the penitentiary in England. We've got a penal reform problem here in Philly. Tell me, how do you find things over there—from the inside?'

He was laughing. He was a very nice man. Quickly, he smoothed it all over and we went on to talk about first impressions of the city and so on, but I couldn't help wondering how many people listened to the programme, and how many newspaper reporters.

Next morning, I didn't have the chance to find out. I was still asleep when the phone rang. It was my agent, Irene, calling from London.

'It's just purely a formality,' she said, 'and there's no reason at all for you to worry, but I think I ought to let you know that there's a warrant out for your arrest.'

The receiver actually fell out of my hand and rolled under the bed. I was crouched on the floor when I picked it up again. Irene was saying, 'Hello—what's happened? Are you all right—?'

'There's a warrant,' I said, 'for my arrest?'

She said again that it was purely a formality, because I hadn't answered to my bail, but she assured me that the Superintendent was being very nice about it, and I wasn't to worry.

I thanked her for letting me know and got back into bed again, where I lay rigidly on my back for some time with my eyes shut.

The phone rang again. It was quite a long time before I picked it up. 'Yes—?'

'Mr Campbell?'

'Yes.'

'You don't know me but my name is Frank Somethingorother and I'm your host for the day. You want to meet me in the lobby in a half hour?'

'Yes. Thank you—'

'How shall I identify you?'

I thought of several ways in which he could identify me. Debauched, terrified, demented. I said, suddenly, 'I look like a mad ferret about nine feet high.'

There was a longish silence. Then he said, 'Would you oblige me by repeating your identification, sir?'

'Tall,' I said. 'Wearing a dark-brown suit.'

'Fine,' he said, reassured. He gave me his identification, which included his weight—one hundred fifty-two pounds.

That day was interminable, and the next and the next. With a warrant out for my arrest I partnered Mrs. Frank Somethingorother at a dinner dance in a hotel. The dance began at six, and we had half a bottle of Californian white wine among five people. With a warrant out for my arrest I sat in the Theatre-in-the-Round, the only person not laughing at the hilarious clowning of Miss Carol Channing. Knowing I going to be arrested at London Airport I sat up all night in the plane that took us home, staring at the back of the seat ahead of me.

We passed over Ireland in the dawn. It looked very green and fresh and innocent. If only I'd never left . . .

There were no policemen to meet me at Heathrow. Only my own solicitor. He looked sad. He said, 'The Superintendent tells me he won't execute the warrant for your arrest provided that you report at the police station at five forty-five tomorrow morning. I think you should do that,' he added, for fear that I might contemplate returning to America. 'Your case comes up at ten.'

I was early for my appointment. It was just after 5.30 a.m. when the door of the cell slammed behind me.

The copy of *Woman's Mirror* was still there.

It felt as if I hadn't really been away.

The subsequent proceedings were shorter than I'd expected, but at least I had correctly estimated their result.

I wouldn't mind going back to Philadelphia some time, just to see what it's like.

What a Bazaar

Now that it's high summer and every green thing droops in the heat we've really got things moving on our plot of ground.

You can scarcely hear the shrilling of the cicadas in the olive trees. The droning of the bees in the lavender is inaudible. Even the thrash of the fire-watching helicopter, as it passes low over-head, has become but a background to the general din.

It all happened very suddenly, as it always does with even the smallest construction scheme in the South of France.

At one moment we had only Manuel and his cement-mixing colleague Saud, putting finishing touches to the little house beside the pool. They were almost entirely silent. From time to time Manuel would call mournfully, 'Saow!' There would be a fairly long interval and then Saud would reply, 'Ouay?' 'Mortier!' from Manuel, and then silence again, broken only by the cicadas and the bees.

But that was before last Monday. Last Monday, at 7.30 a.m., Gaby and Mario arrived, and *their* cement-mixing colleague Hassim. Gaby arrived in his car and Mario and Hassim on their motor-bicycles. They distributed their machinery in various shady spots about the property. They greeted Manuel and Saud with rather curt nods. No équipe around here likes to work in the presence of another. Each équipe believes that the other is doing their job all wrong, so that everyone gets a little nervous.

Gaby and Mario, however, began to lay their paving beside

the pool while in the background Hassim and Saud mixed their cement in separate piles.

Then the Carreleur arrived. He's the tile-layer, a tall, shy and silent young man who wears a girl's straw hat. He doesn't have a cement-mixing friend, so he began to mix his own, well away from Saud and Hassim. He had left his car outside the gap in the fence.

Whilst we were getting a little crowded, things were moving ahead peacefully enough. Then eight men from the Electricité de France arrived, towing on a low-loader a concrete pylon perhaps fifty feet long. The Carreleur's car was in the way so that had to be shifted and then the Electricité équipe, with two pneumatic drills, began to dig a hole on the corner of the road in which to plant their new pylon. For this, of course, the electricity had to be cut, and this put our pump out of action so that there was no water in the house.

Already the cicadas, the bees and Manuel, Saud, Gaby, Mario and Hassim were raising their voices a little, to communicate above the noise of the drills. It wasn't anything serious, however. No more than, say, Oxford Circus at midday.

Then, for the first time, the sprinkler system équipe arrived— seven men with picks and shovels, led by a young man with a beard. Four of these workers appeared to be called Hassim and the other three Abdallah. They began, with incredible speed, to dig deep trenches up the centre path, across the grass and through the shrubbery. The whole garden erupted suddenly into long, high mounds of yellow clay. The various Hassims and Abdallahs encouraged one another with shrill Arabic cries. While this was going on the bulldozer arrived, to dig the hole to store the water which will eventually serve the sprinkler system.

The Electricité de France men had to move the low-loaded pylon to let le booldozaire in, while the postman in his yellow van blew his horn continuously, in an effort to get past with the mails.

We now had at work Manuel and Saud, Gaby, Mario and

their Hassim, the four other Hassims and the three Abdallahs, the young man with the beard, the Carreleur and the driver of the booldozaire, who had a Saud and a Hassim of his own. Into this fairly crowded scene, then, walked Marcel, our handyman. I'd forgotten it was Monday, his working day, and had some difficulty in finding him a place in which to work. He eventually set to to level a patch of ground between three trenches and the bulldozer hole, with the intention of planting some dwarf beans he'd brought with him.

At 9 a.m. our splendid *bonne*, Madame T., arrived. '*Quel bazar*', she observed briefly, having reference to the activities in the garden, and—seeing that there was no water—began work on the dining-room tiles with the electric polisher.

Madame and I breakfasted gently in the little inner courtyard. Three people were coming to lunch and as yet there was nothing for them to eat and thanks to the power cut the icebox was melting.

We toasted ourselves in the sunshine, utterly at peace with the world.

A Rough Old General

The weather here has been patchy in the extreme for the French General Strike.

Round about now we should be sitting in the shade, pointing out to one another the enervating quality of the heat and the exceptional dryness of the ground, and advancing from there into a comprehensive destruction of the character of the Mayor, who year after year fails to provide us with sufficient water for our vegetables, but in fact—ever since the students started things in Paris—we have been subject to weather conditions not dissimilar to Wigan in August.

That is, a more or less continuous overcast, leading to sudden and torrential rain.

It has put us out. The General Strike, we tell one another, would be just tolerable if the sun was blazing down as it should do at this time of year, and we were able to complain with fire and passion about the shortage of water, but as things stand we've got nothing concrete to level our indignation against.

Round about the time that the workers took over the Renault factory two parties developed in the village. One held that the almost continuous rain was good for the earth; the other maintained that the tender young roots of the tomato plants were being drowned in an excess of unseasonable moisture.

Suddenly, however, the sun would come out, the ground would begin to steam, we'd start to couple up our hoses, and then the rain would come back again. It came back, in fact, on the same day that President de Gaulle returned from Rumania, leaving us with so much to talk about that we could scarcely apply our minds to anything specific.

Once, we talked for perhaps three hours about a Spanish workman who had been seen buying ten kilos of flour in Pré du Lac, while his wife, in the establishment next door, was stocking up with an equal quantity of sugar. We agreed, fairly early on, that it was only the foreigners who had panicked, but that it was only to be expected of them, because of their invincible ignorance, and then we said the same thing, comfortably, over and over again.

Of course, it wasn't all as easy as this. One or two of the older ladies got a touch of the cafard, rather in the way of hens getting their back feathers ruffled by an unexpectedly chilly puff of wind. Madame T., for instance, said she couldn't bear to think of the noise being made in Paris by students tipping over motorcars, and many of us shuddered in sympathy. We were all relieved by her decision to walk down the hill and to take a cup of coffee with her sister-in-law, as an antidote to this *crise de nerf*. She was certainly easier in herself when she came back again, incidentally letting slip the information—provided by a friend of her sister-in-law's cousin—that row upon row of shelves in the Supermarché in Cannes were standing empty, as a result of various urban elements hiring whole vans to drive away perhaps a thousand tins of sardines per head. In full, plenary session we agreed that an occasional sardine was quite interesting and tasty, but that too many of them would be bad for the liver. That night a policeman was shot dead in Lyons, and the older ladies were driven to making whole batches of fresh ravioli, to keep a return of the cafard at bay.

We have, of course, had no post for more than a week, while at the same time seeing a great deal of the postman. The heavy rain has promoted a tremendous harvest of snails in the long grass bordering the lanes around here, and the postman—

a young and eager lad—hunts them all day long in his large, black DS.19. Perhaps his official position gives him a private supply of snail-hunting petrol, because petrol is getting a little short.

We held many discussions on this subject. In view of the fact that the Customs are on strike at the frontier post in Menton, some people hold that it would be worth while to drive into Italy and to bring back enormous quantities of loot, including several hipbaths full of petrol, but then the other side maintains that while we were queueing to get over the border the strike might be settled, and we would have missed lunch to no purpose. It's very difficult.

The telephone has also become uncertain. The exchange has a tendency to say that you can only ring a doctor, so that we all ring one of the three doctors and tell him to ring our friends and ask them to come to dinner on Tuesday, and *merci mille fois, M'sieu le Docteur.*

It is, in fact, now Tuesday morning. Miss Beryl Gray, that distinguished ballet dancer, hopes to leave Nice Airport for London this very afternoon. I am entrusting this message to her, in the hope that she will be able to take off.

In the meantime I think the sun is struggling through, and the sea is already beautifully warm.

It's certainly been a rough old General Strike down here.

Knickers on the Grass

MANY people must be familiar with the aftermath of having a few small jobs done around the house and the garden by Fred, Charlie and Len.

Some light carpentry, a touch of paint here and there, an adjustment to the drains—something like three weeks work for Charlie, because he's the one who does it.

Fred is the one who handles the administrative side, and the paper work. He does the planning and gives the orders. 'Start here, Charlie, go down to here on the shovel and I'll tell you when to stop.' The paper work is concerned with the lunch-time racing edition, and this is where Len comes into his own.

For much of the time he holds a screwdriver in case Charlie, digging up to his knees in mud, might have need of it, but he's got another important task—that of nipping round the corner to the betting shop in case Fred is suddenly inspired by his study of the newspaper to make yet another investment. Len often takes the screwdriver with him on these occasions, and leaves it there, and has to go back and get it, if it hasn't already been nicked.

There's nothing very surprising in the smooth way that all three of them work together, but the thing that does astonish the employer is the huge variety and ample substance of the debris they leave behind.

Folded neatly into a fork of the apple tree are a large number

of racing editions, sodden and pulpy with rain. In the bathroom, which has not been painted, is a tidy pile of eight empty
tins of paint. In the potting shed there's a pair of broken shoes,
without laces. An old waistcoat hangs on a nail behind the
door. In the house itself a pile of shavings has been swept into
the pantry and on the windowsill of the spare bedroom is a
paper bag containing quite a number of rusting nails.

But this is British rubbish, familiar stuff that even one handyman after one day's work can easily leave behind. It often takes
less than twenty-fours to clear away.

Contemplate, however, if you will, the veritable mountains
of *ordures* that can be deposited upon half-an-acre by a team of
about fourteen French, Italian, Spanish and Algerian workers,
toiling away from dawn until dusk and—a vital factor—lunch-
ing on the site.

On Wednesday morning, for instance, after a quick 5.30 a.m.
turn around the property, I counted nine litre bottles of
various lemonades, orangeades, beers, wines and *syrops*, all
with violent labels in primary colours, lying everywhere on
what used to be the grass.

There were three yellowing copies of *Nice Matin*, folded into
admirals' hats against the sun, carefully stashed away under
heavy stones. One of them read: 'A Virgin of Thirteen Years
Sets Fire to Herself after her Lover has Flown Away.'

There was a cardboard box labelled 'JAVEL-BEC 30
DOSES' which seemed to have contained a lot of disintegrating
tomatoes; a fearful brimless felt hat; a pair of oil-stained,
pink, female knickers lying beside the bulldozer; a tennis
shoe without a sole; a foot of salami convered with ants and
everywhere, plastered against the trees and bushes by the wind,
were cheese papers and ham papers and bread papers and the
remnants of paper sacks of cement.

Without showing it, of course, I was well pleased. In an
English garden it is the accepted thing for master and man to
work together. The man, indeed, often shows a quiet pride in

the competence of his master; 'His Grace is a dab hand at his
florabunda.' But here in the South of France the division between the two is absolute. The master isn't allowed to do a
hand's turn and if he attempts it the man stops whatever he's
doing and comes over and watches and shakes his head and
tells you you've got to have the habitude for that sort of thing,
even if it's a little light raking of a patch of gravel. Then he
does it himself.

For the last couple of weeks, therefore, with all these French,
Spanish, Italian and Algerian experts around the place, the
most manly task I've been able to find is polishing the plate of
the sundial, which still hasn't got a base and probably won't
work in this latitude anyway. I could scarcely wait for everyone
to go, to start clearing away the bottles and the salami and the
old felt hat.

Most of them went the day before yesterday. They took
every crumb of their litter with them, including the bulldozer's pink knickers.

Obviously, it's going to take some time to become accustomed
to the ways of this foreign country.

A Rather Better Hole

'ANOTHER filthy morning on the ghastly Côte d'Azur.'

Such is now my buoyant announcement to Madame at the start of each new day, as tenderly I lay the pot of China tea by her bedside.

At 7 a.m. the sun is already hot and golden. The sky above the mountain behind us is shimmering mother-of-pearl, but the deep blue that will last for the rest of the day is already breaking through.

We refer to these conditions as filthy out of deference to more experienced persons who tell us that the weather is so much more reliable, darling, in the Bahamas, Marrakesh, Sardinia or Corfu.

Certainly, before November, when it tends to become just a little bit nippy, we shall have a number of thunderstorms, with drenching rain to revitalise the garden, but by and large there are going to be about twelve hours of sunshine every day until then.

Tolerable conditions, if one does not demand the earth.

It is also out of deference to those who prefer, darling, the Bahamas, Marrakesh or Marbella that we refer to the 'ghastly' Côte d'Azur.

For the Beautiful People the Côte is indeed finished, with German coach tours sightseeing in the Casino in Monte Carlo and even middle-class French families infesting the beaches of Cannes. No wonder they have retreated to the Costa Smeralda, where they can be accosted only by the Prince and the Princess, the Duke and the Duchess, the Count and the Countess and, with any reasonable luck, the Aga Khan.

We find it, however, convenient to be within twenty minutes of the coast, of the airport, of Cannes, Nice and Antibes. It is agreeable to come down from the mountains, from the silence of the centuries-old olive trees, to see what's doing in Juan-les-Pins, and to retire as quickly as possible when we see what is.

But, now that we have finally emigrated, the real joy is to be living on a Continent, to be able to get into the car and drive to Oslo or Lisbon or Istanbul if we have the mind, and there's no need to book a ticket to get out.

I lived in Ireland on and off for thirty years, and for another twenty in England, and I believe I always had the feeling at the back of my mind of being trapped. Aer Lingus regret to announce that due to fog all flights are cancelled until further notice, or owing to a technical hitch among the baggage porters all B.E.A. planes will be grounded until the end of time. These mishaps would naturally coincide with hurricanes in the Channel or the Irish Sea, so that all shipping would be confined to port until even one's desire for escape had gone.

But here, only the other day, we had a long conference under the orange tree in the garden about whether or not we should drive quickly to Venice, to have a look round for a couple of days before the tourists got there, but decided against it in the end. The orange tree is in full perfumed bloom, alive with bees, and the falling white petals litter the small paved garden. They look untidy, like bread-crumbs, and it takes us a leisurely half-hour, with intervals for a glass of wine, to sweep them up. We did that, instead of going to Venice. Very pleasant indeed.

'But what do you *do* down there all day long?' people in London want to know, as they pursue their full, rich lives looking for a parking space, commuting to Ponder's End and getting three wrong numbers in succession on the office telephone.

The truth of the matter is, as most people discover who have escaped from cold, Northern cities into a sunlit countryside, that the day simply isn't long enough for all the totally absorbing pottering about that one has to do. Even the shopping is a pleasure. The warmth of Madame André's greeting never fails from day to day. Every vegetable, each single apple, is chosen and discussed with loving care. And a litre of wine is 2/6d.

'But surely,' the people in London say, still mesmerised by last night's telly and eagerly waiting for this evening's unchanging dose, 'surely your minds will decay. You'll go to seed. You'll simply turn into vegetables.'

We assure them there is no danger of any kind. When they cannot be avoided, because they arrive on your doorstep, the English residents around here are capable of providing drama even up to the high standards of the late Lytton Strachey.

I should like to go into this, circumspectly, at a later date.

For Clover's Sake

THE sun rose properly the other morning, a golden fiery red, unencumbered by the Welsh valley, slagheap-coloured overcast which has now prevailed in these parts for as long as anyone can remember.

I knew it was up properly because it pierced the ventilation hole in the shutter, bored its way right through the curtains and laid a small, scarlet ball of heat on the arm of the sofa.

It convinced me that something had happened to General de Gaulle.

The belief is sincerely held around here that the General, while not being perhaps 100% responsible for the weather, at least has cast a shadow upon it as large and as long as himself. We have jokes like, 'During me the deluge'. and many agriculturalists cannot wait for the General Election to reach its conclusion, so that the General may retire with honour, and the sun come out to sweeten the vines.

For myself, I'm having clover trouble. Early in the year, before the rains came, what was said to be 'miniature' clover was planted between the flag stones at the other end of the pool—little threads of green to break the glare of the summer sun. Then the monsoon set in for ever, the clover began to grow and almost instantly—it seemed—was about a foot high and threatening to invade the whole property. I cut it down from time to time with the shears, finishing off with a pair of scissors, but the process left tons of clover to be swept up and by nightfall it was back again just as tall as ever.

On this particular morning, then, I was lying in bed, watching the marvellous ball of heat moving along the back of the sofa, when I thought of a rabbit. Set a rabbit to work on the clover and in no time it would be down to the bone. But what would happen if the rabbit got tired of clever and insisted upon lettuce? The labour of providing it—and the expense!

At that moment the shepherd went by on his motorised bicycle. I knew who it was because of the cacophony of barking that accompanied him. He's got two dogs. One of them looks like a mad, black bear with the legs of a shire horse; the other is a kind of very young Alsatian, but with the almost invisible tail of a boxer. As they bound along beside the shepherd's motorbicycle the Alsatian tries to eat the bear-like thing, urging himself on with hysterical yelpings. The shepherd pays neither of them the least attention. But his passage started me thinking about sheep.

'Good morning, shepherd. Can I borrow one of your muttons?'

('*Je voudrais bien emprunter un de vos moutons*'—?)

But which one of the sixty-odd that he drives past the gate every day? And would I have both the brutal dogs as well to keep it in order?

Smoothly, then, as a hot knife through butter I solved the problem. I got up, put on a pair of shorts and a thick, towelling dressing-gown, went out to the pool and began to browse upon the clover myself—not eating it, of course, but pulling it up by hand, putting the residue into a bucket. The system worked perfectly, on this still, golden morning. No clover to be swept up afterwards, and the skirts of the thick dressing-gown protecting the knees from the abrasion of the stones.

I entered upon a state of bliss, greatly increased—shortly afterwards—by seeing Madame at the other end of the pool in a very short nightdress, sweeping fallen olive blossom away from the water.

'Good morning,' I said. 'Do you know what time it is?'

'It's heaven,' she said. 'It's ten to five.'

'If we were in London,' I said, 'and awake, we could only be coming home from somewhere where we'd stayed much too long.'

'Or,' she said, 'you might be in Chelsea police station, turning out your pockets at the request of the sergeant in charge.'

We had breakfast in the garden at 5.15, and the first little sip of Cinzano and Campari at 9 a.m.—a justifiable indulgence in view of the fact that we'd been up for more than four hours or, by London standards, having already got as far as lunch. The rest of the day was endless, and a bright gold.

Selective Employment Tax, Purchase Tax and multiple telephone directories had never seemed further away. Nor, indeed, did the General.

He was back again next day, however, soaking the whole of the Alpes Maritimes.

Votez Communiste! Or any other way you like. Just let's get it over, for the sake of the clover.

The Huge Thin Freeze

You may have had a cold snap in Britain over the New Year, with one or two attendant inconveniences, but it would have been almost unnoticeable in comparison with the full-scale disaster that fell upon us here in the lower Maritime Alps.

At least an inch of snow all over everything on New Year's Eve! Some of the older people in the village said that their grandfathers had often spoken of something similar when they were young, but it certainly hadn't happened since.

It found us unprepared, particularly in the way of protective clothing. Normally, in these parts, there is no call at all for wellingtons, anoraks, fur hats or mackintosh trousers, so that during the morning there were a number of unusual ensembles to be seen, for those who had to move abroad. One old gentleman, eager to save his ripening olives from the blizzard, was seen to be doing so with the arms of a child's jersey tied over his cap, a heavy red curtain round his shoulders and a skirt or some plastic material wrapped round his waist with a length of rope. To protect himself from the snow falling from the trees he had a parasol of old fashioned make.

There was another unusual feature of the morning. In the ordinary way, on New Year's Eve, the surrounding woods would have been filled with the sound of shot and shell, as the hunters made their clearance of everything that moved, including little Robin Redbreast, but this time the silence of the grave lay over all—until, that is, the motorists began to stir, and found they couldn't.

Unfamiliar with snow and ice they dealt with it by engaging bottom gear and pressing the accelerator down to the floor. From all over the valley came the roar of engines in torment, as though some gigantic stock-car race was in progress, or about to start, for no one seemed to move.

I, of course, a motorist from northern climes, was much more skilful. I got out the chains which had incredibly enough come with the car when I bought it secondhand and tried for nearly ten minutes to put one of them on. At the end of that time it became evident that I didn't know how to do it or that they weren't wheel-chains at all, but were machinery intended for quite some other purpose. I put them back on the nail and started off slowly downhill to the village.

I nearly had an accident at the first corner. A man, revving his engine wildly, was grinding slowly uphill out of his garage, threatening to cut across right in front of me. At the last moment, however, he lost his forward momentum and began to slide back again, the back wheels a blur. He waved me a sad farewell as he disappeared backwards into the garage he had recently left. A crash indicated that he'd been stopped, probably by the rear wall.

Round the next corner I missed, by an inch, a man boiling up the hill in a sports car with skis on the roof. As he went past he roared, '*Priorité!*' Obviously an Alpine expert, he had given me a useful tip—namely, that those coming up had the right of way, even if you couldn't see them.

I met him again, when I was coming up and he was coming down, and only missed him by driving into someone else's garage, fortunately empty at the time. I sprang out, to bawl '*Priorité!*' after him, and performed the perfect parabola that can be done only on ice. That is the feet out from under, in the air for a moment and then the whole carcase striking the ground with a shuddering crash.

After some time I got into the car again and found that while I could back it out of the garage I couldn't get it going again up the hill so I put it back in the garage, off the road, and walked up the hill to our house to get a bucket of sand. I came back with the bucket of sand and a huge sheet of three-ply wood, to find five cars stranded at various angles up the hill, headed by a large Ford Taunus, which wanted to get into the garage now occupied by my own vehicle. The shock caused my feet to leave the ground again and this time I got a kind of planing effect from the sheet of three-ply so that I seemed to gain extra altitude before striking the ground.

The sun, of course, came out shortly afterwards and melted all the muck in a trice and we had drinks in the garden but all the same it will be a long time before I forget the Big Freeze in Alpes Maritimes on the last day of 1968.

Mr. Bulldozer

MONDAYS are the frantic days here because on Monday Marcel, the handyman, arrives and I've got to find something for him to do before he finds it for himself.

We used to see a little of Marcel around the village, before he came on the payroll, and the little we did see of him we saw often. Just his head and shoulders, appearing out of holes in the road—an unnerving sight because of his round, protruberant yellow eyes and his long, drooping moustache of the same colour, added to small, pointed teeth like a cat and a permanent, beaming grin.

Marcel used to do a great deal of free-lance hole digging in the roads around here, and seemed always to be about six feet down. Part of his equipment was two large steel sheets, which he would manoeuvre into position from below to permit the passage of traffic across his trench, and then the head, the yellow eyes, the pointed teeth and the beaming grin would pop up in an empty space between the steel sheets, and Marcel would greet the world: from his worm's eye view.

Due to the size of his smile and the sharpness of his teeth he often seemed to be on the verge of snapping at the tyres as we drove past, but he never did.

In two or three years all that we saw of Marcel was his head and shoulders, so that it was fairly difficult to estimate what the rest of him was like. Then, as I say, he came on the payroll, and we saw it.

He clocked in exactly at 8 a.m. for his first day's work.

It was already very hot, so that he wore what looked like the whole of one copy of *Nice Matin* folded into a paper hat. It came very low down on his head, so that from the front really only the pointed teeth and the long yellow moustache were visible. He was bare to the waist. Some way below that area a pair of striped bathing-trunks appeared, and disappeared almost instantly. Further down his evidently bare feet were encased in battered, brown working boots the size and shape of rather large, plump cushions. The overall dimensions of Marcel were comparable to those of about five and a half feet of the trunk of a fully grown tree.

We shook hands. The pain shot right up my arm and rattled a few vertebrae together in my back.

I showed him his first task—the clearance of the jungle of weeds that had grown up inside the fence surrounding the modest property, a dense and lethal mass of brambles and other spikey horrors that grow so quickly in the heat that they wrap themselves round your legs, unless you keep moving. I then presented him with the tools for the job, some rather light trowels and forks and things purchased in the Monoprix. Marcel took them eagerly. In his hands they looked like crochet hooks. He withdrew to the furthest corner of the property and then, in the way of the men of Provence when they are faced with imminent work, he sat down under an olive tree and lit a cigarette.

I left him to it—after all, it was exceptionally hot—and we went down to the village to get the lunch. When we got back Marcel was clearing the top of the wall that divides our bit from Andre's, having already cleared not only the whole of the inside of the fence along the road but the outside as well. His smile was as big as a shark's, but much, much more friendly.

That's what makes Monday the frantic day round here, the knowledge—from 5.30 a.m. onwards—that today is Marcel's day for pottering in the garden, and that if I'm not there at eight to greet him with a working plan he may well potter up all the olive trees and replant them in a more pleasing design.

He has already dug what will be the vegetable garden to the depth of two feet, despite the fact that the ground is yellow clay, baked by the sun to the consistency of reinforced concrete. Perhaps I can get him to level it and then dig it up all over again.

It's just a matter of pushing the right button and this, of course, is the source of my unease. It would be so much more peaceful if we had a lovely, sodden English garden and a handyman called Fred, whose rheumatics alone justify the National Health.

Fred would arrive at ten, speak about his rheumatics until eleven, re-pot a plant in the safety of the greenhouse and go home. Usually, however, we have to slow Marcel down round about six in the evening with a pint glass of wine, and even then he's reluctant to halt his bulldozing.

The pressures of life down here will, no doubt, age me before my time.

A Feast of Late Appalling News

EVERY evening at about five o'clock I drive down the hill to get the papers, or the paper, or none at all, because they haven't arrived.

Or, perhaps, it's raining and I decide not to bother until tomorrow, when the third possibility—the none-at-all—will surely have been removed and with it a temporary but nonetheless urgent sense of fear.

In the old days in London I used to be stationed behind the front door every morning at 7.30 a.m. waiting for the delivery child to ease them to me through the letterbox. When, as happened several times a month, there was a new and untutored child who delivered them to someone else I used to go half-mad with fury and ring the newsagent and shout at him until he got tired of it and rang off.

Here, in France, the delivery system is different. I believe, while being dependent upon hearsay, that the English newspapers of that day arrive at Nice Airport some time during the morning and are then transported by bus to the newsagent in Grasse, who re-ships them some time during the afternoon to the bar in the village.

Sometimes, when I've got nothing else to do, I sit trembling at the prospect of the load of appalling news that will reach me by this devious route about five o'clock this evening—if, that is, it arrives at all.

Unfortunately, I still regard British news as being the most real, but this is not due, I hope, to the last vestiges of insularity.

F

French news, as contained in *Nice Matin*, is obtainable at our
local bar, often with a cognac as an antidote, about 8 a.m., but
it never seems to me to be as punchy, as immediate, as the
British kind. There's a lot about heart transplants, minor
robberies and fearful crashings on the national routes, together
with interminable and totally unreadable accounts about the
very rigourous steps that the Government is taking to support
the economy, but you never seem to get anything about
Harold Wilson looking unusually glum with a heavy cold or
actresses found unclothed under the billiard table of the Duke
of Mud.

Nice Matin is, in fact, a bit local for me. Consequently, when
I collected no less than five English papers the other evening,
after an unendurable period of none-at-all, it was just as
though I had come upon a forgotten store of LSD stashed
away against a rainy day in the turn-ups of a pair of trousers,
placed for safety's sake at the back of the refrigerator.

(Note: One keeps up diligently with the English way of
life even down here.)

Not to spoil this feast of appalling news I didn't even glance
at the headlines but hurried home with all speed, only to find
that several other emigrés had dropped in for the hour of the
cocktail—by which I mean that they were looking for whisky,
which has now shamefully rocketed to about the same price as
it is in England.

'Oh, good,' they cried, 'newspapers!'—and fell upon my
stock just as though it were fine old malt. Not being professional
newspaper readers—there were women among them—they
grabbed the first one that came to hand, leaving me with the
most recent, dated December 10th.

It was like trying to take a trip on powdered aspirin, after
someone's nicked your own carefully prepared sugar cube.
Like no bite to it, man.

I read of a 30% increase in the emoluments of all nationalised
industry chiefs, when the last I'd heard from the industrial

scene was an offer from Barbara Castle practically to do a strip in Transport House, in exchange for an end to wage demands.

Sir Leslie O'Brien was back—*again*—from Basle, saying he hadn't asked the Central Bankers for a tosser, when only last Friday it had seemed that Britain was about to be finally bankrupted by a Coalition Government, with Jeremy Thorpe as Chancellor of the Exchequer.

Last Friday I had read, with bitter tears, that noble Tommy Docherty had abandoned British soccer and was taking up a job in Spain on New Year's Day. Now, however, it seemed that he'd forgotten all about that and was dickering with Aston Villa.

I couldn't imagine how everything had turned so terribly back to front, in so short a space of time and now—when the others had finished with my newspapers—it didn't seem worth while finding out. Obviously, if one didn't keep with it, it never happened at all.

(Memo: Buy no newspapers between December 20th. and December 31st. That will effectively cancel out the horrors of you-know-what.)

Santa from the Sahara

As a special bonus for Christmas we have, condemned to imprisonment in the cellar, three bottles of absolutely undrinkable Scotch whisky—as heavy a burden as anyone could wish for in this season of peace on earth and goodwill toward all men.

Of course, it's not true to say that it is 'absolutely' undrinkable. The first guest we'd tried it on got as much as half of it down before his lips began to move uneasily, and the questioning look to come into his eye. Even at that he didn't blame it on the whisky. He said he was awfully sorry but he thought that a drop or two of something else—perhaps something like paraffin—must have got into his glass and could he possibly jettison this lot and have another one?

That's when I had to tell him, sombrely, that it was all the same, but nonetheless that I did not share his view about the paraffin flavour. 'We find it,' I said, 'a little heavier than paraffin. More like groundnut oil mixed with diesel fuel and perhaps some otherwise inedible herb from the Middle East. A difficult flavour to assimilate, but perhaps after another one—?'

He said he thought not, thereby condemning the remainder of the beverage to the cellar, because if this particular man cannot drink it no one can.

When this kind of disaster occurs it is impossible even for the most generously-minded of us not to seek around to apportion the blame and—after less than a second's reflection—

I find I can lay it squarely upon world conditions, with particular reference to the present disordered state of European
currencies, and international trade.

If, for instance, General de Gaule had not taken his arrogant
and crazily optimistic stand about the value of the franc we
should not have been skulking about the back streets in Cannes,
looking for a bob or two off the staple necessities of Christmas
cheer. Thanks to Cohn-Bendit and the Renault workers and
the General's panic concessions to wage demands the price of
a bottle of Scotch is about the same as it is in England or even
Aberdeen, when before the revolution it was a great deal less.
When, furthermore, you realise that you are paying in devalued
pounds the back streets and the cut-price boozers become stern
economic sense.

We found a lovely one behind the Old Port. Father Christmas
was even present, in the form of an Algerian youth of about
eighteen with rather sparse cottonwool whiskers. He was
leaning against a car, smoking a cigarette, and chatting in
Arabic with some friends in mufti. It seemed probable that he
was also officiating on behalf of the horse butcher next door,
since the French tend to save on the festive appearance of
Christmas and to concentrate on the increased trade.

At the saving of about 6/- a bottle we bought three of vodka,
of brandy and of pastis—all well-known and reliable brands.
Then we tripped over this unfortunate whisky, the name of
which it would be libellous to disclose. It retailed at 39/6d., but
it looked all right with 'Fine Superior Old Scotch' appearing
twice on the label and in the middle some equally reassuring
message in Olde Englishe script, but just too small to read
without the aid of glasses.

I read it, with glasses, when I got the bottles home. It said,
in Fyne Olde Englishe script, 'Perfection Scotch whisky
blended from Scotch Highland malts and grains produced by
world famous Scottish distillers under British Government

F*

supervision' and I very nearly smashed the bottle to the ground there and then.

'Perfection Scotch whisky' indeed! It wasn't even English— or Scottish. It could have been written by the Algerian Father Christmas. And then to have the British Government—those guilty, frightened men—lurching in on the act, to supervise the production of world famous Scottish distillers, who weren't even named, because they couldn't be because they were undoubtedly Wee Jock and Big Rob, boiling it all up in the sink in some Glasgow slum. . . .

But this is the season of peace on earth and goodwill toward all men. Let me wish you all a very happy Christmas and a merry end to a ghastly old year.

For ourselves, we may be short of fine old Superior Scotch in the cellar but we've certainly got the makings of a barrage of Molotov Cocktails, which may well come in useful in shining 1969.

The Laying to Rest of Mug

To:
Malcolm Muggeridge, Esq.
Snuffley's Bottom.
Under Whichwood.
Fuffex—
Or wherever you are.
Dear Malcolm,
    It was a sensuous pleasure—if you will permit so lewd
a confession—to be mixing pure white paint the other morning
in the new little house beside the pool.

For the last three or four days we had been having heavy
cloud, thunderstorms and sudden torrential rain, but then the
mistral arrived from the north-west and blew all this muck
back into Italy, where it belongs, leaving us at 5 a.m. with a
pellucid sky washed clean of all foreign bodies.

The new little house faces east, so that I was able to get the
full benefit of the dawn and the round yellow sun so close that
I could almost touch it. It sparkled on the pool and danced on
the ceiling, which will be even whiter after the second coat.
The cicadas creaked away like mad in the olive trees. The
large lady toad, who keeps falling into the pool and being
rescued, sat plumply on the honey-coloured stones in front of
the summer house, waiting for one of the huge sapphire dragon-
flies to make an error in its flight.

Everywhere there was light and colour and—after the rain—
a bounding exuberance in every growing thing. The whole

world seemed to capering for joy, like the light on the ceiling, and then I saw I was mixing paint on your face.

I'd put this sheet of newspaper on the floor to protect the new tiles, which are going to be honey-coloured too when when they've been polished, and there, right beside the paint tin, was your photograph, under the heading, 'Why I Love England'. Your mouth was slightly open and you were looking up, with proper apprehension, at a photograph in the next column of lowering black Fuffex clouds, about to empty their lot into a patch of storm-tossed, sodden bullrushes. You looked jolly healthy, still handsomely sun-tanned after all the months you used to spend down here at Roquebrune, right beside the wine-dark Mediterranean.

Glad to see that you were still churning out what S. J. Perelman calls 'merchandisable threads', I was about to resume my paint-mixing when a word or two of your sermon caught my attention. 'The country-side,' you wrote, 'which I have always defended, continues to delight me with its variations.'

I actually stopped paint mixing. By all accounts—many of them provided first-hand by pallid refugees arriving at Nice Airport—it has been raining in England and presumably upon Snuffley's Bottom without cessation of any kind for at least three weeks. Not much variation of climate there. You'd got it so wrong, in fact, that I wondered if you might not be back again in the South of France, rather desperately spinning merchandisable threads about the sodden homeland from here.

I read on, concerned for your problem, knowing it only too well myself, only to come to the conclusion that you must indeed be in Snuffley's Bottom, with the mud up to the top of your little rubber boots, because soon afterwards, through the paint, I was able to decipher the observation, 'I think with pity of those expatriates who, to escape income tax, drag out their lives on the Riviera . . . what miserable existence . . . picking up the English papers as soon as they arrive and even—

so hard-pushed—actually reading the obituaries with a kind of sick longing! After all, the dead will be laid to rest in England . . .'

At this point, unfortunately, the lady toad fell into the pool again and I had to fish her out and dry her with the paint cloth and set her up once more in dragonfly-watching position. This small mishap broke my concentration, so I didn't finish reading your piece, and in any case it was getting so hot it was time to slip into the pool, to drag out my life on the Riviera as best I could.

But I'm still worried about you, because you're a very nice man. I mean, this thing about being laid to rest in England. There's a charming little cemetery, quite close to here, in a ring of mimosa trees that are covered with perfumed yellow blossom almost throughout the year. This is for me. Can I book you a neighbouring plot?

At least it will be more comfortable than Snuffley's Bottom, in that they won't have to pump out the rainwater before lowering you into the hole.

Looking forward to your esteemed order, I remain,

Yours, etc. . . . .

Nero, Rufus and all that Mob

Mrs. Siddons, a highly theatrical young poodle who lives on the other side of the road, usually begins it.

Either because she suddenly remembers she hasn't done it for some time or because she sees a leaf in some unusual, even monstrous, colours, she makes with the barking.

It goes off as abruptly as a pistol shot, and is infinitely more destructive to the nerves.

Mrs. Siddons, neurotically distraught, suddenly goes, 'Wowowowowow!' It is, simultaneously, a cry of fear and a yell for help, and it has the same effect upon the eardrums as transfixing them with two steel knitting needles, which meet in the middle in one of the more delicate lobes of the brain.

It must be allowed, on the other hand, that Mrs. Siddons is a short-time hysteric. One good, 'Wowowowowow!' will often last her for as much as twenty minutes. Unhappily, however, it has the effect of setting off Rover, round the corner and some way down the road.

Rover is some dog. Standing on his back legs, as he invariably is, maniacally clawing at the wire fence that contains him, Rover would be about seven feet high. That part of his ancestry which wasn't provided by a sheep seems to stem from the horse. The canine ingredients in his make-up are more or less limited to his bark, but that's all dog.

Set off by Mrs. Siddons, Rover goes, 'Warf-warf-warf-warf-warf!' by the hour. There's a lot of her hysteria in his delivery.

It's got the same frantic quality, but tends towards the baritone rather than the soprano.

There is only one thing that stops Rover going, 'Warf-warf-warf-warf!' and that's when he suddenly gets a 'Woof!' in the middle of it. This 'Woof!' strikes him as unexpectedly as a hiccup, and renders him silent for almost thirty seconds.

It's quite a sight to see the great shaggy black face of Rover dripping above the top of the fence, with the mad, sand-coloured eyes almost crossed in concentration, as he tries to work out why this 'Woof!' has so dramatically interrupted his warfing. But then Mrs. Siddons up the road sees the top half of a worm, or a cloud or pretty well anything, and she yells 'Wowowowow!' and that gets good old Rover going again and he warfs away like a maniac until once again he runs into another 'Woof!'

Sometimes, of course, it happens that the thirty seconds of silence which follow coincide with an equal period of silence from Mrs. Siddons, but good old Rufus is there to take up the slack.

No one knows to whom Rufus belongs. He's thick-set and red and woolly, and there seem to be a number of badger genes there in addition to a sprinkling of cocker spaniel. He spends his life trotting purposefully up and down the road and going 'Wurf!', once perhaps every three yards.

One almost gets used to Mrs. Siddons. The noise made by Rover is so deafening that it numbs the senses. But the sight of good old Rufus bumbling along about his non-existent business, going 'Wurf!', trot-trot-trot. 'Wurf!' trot-trot-trot, 'Wurf!' unhinges the mind. ·

Particularly when he runs into Nero, the Alsatian Boxer or, equally fairly, the Boxer Alsatian. Nero is a young dog of blameless disposition, who spends his time trying to make friends with every one, every dog and, indeed, every thing. He is assisted in this fairly laudable enterprise by being almost able to speak. That is, he goes, fortissimo, 'Wuh-ho! Wuh-ho!', addressing himself impartially to people, hens, motor-cars, cyclists, prams and other dogs. He smiles as he goes, 'Whu-ho!' He might well be saying, 'Huh-lo!' if only he could handle the aspirate, but he shouts too loudly for that.

He unsettles good old Rufus. Rufus is too busy to be propositioned. Faced with beaming Nero, Rufus strings his normally divided 'Wurfs!' together into a cascade of 'Wurf-wurf-wurf-wurf', so that it sounds like a stick being rattled along wooden railings. And that starts Mrs. Siddons off and Rover is still at it anyway and Nero is bawling 'Hello-hello-hello-hello!' and the extraordinary thing is that none of the owners of these deafening and discordant beasts seem to notice the pandemonium they create.

The owners talk to us in their normal voices with the very air being rent by wurfs and warfs and wowowowows. From time to time they even smile upon their yelping and revolting pets.

With so many dogs around it's no wonder the world's in a state of jitters.

The Rise and Fall of Zizi

ONCE upon a time, in a hill village in Provence, there lived a beautiful little girl called Zizi.

At ten years of age Zizi's hair flowed down almost to her waist. It was of a unique colour—a kind of bronze that the sun flecked with gold in summer, to match the yellow of her cat-like eyes. Zizi had another unique attribute—or at least pretty unique for the region. She had a short little body and very long legs, instead of the other way around.

At ten years of age Zizi looked, in every way, to be about fourteen but she was liked even by the younger wives in the village. She was gay and mischievous and innocent all at the same time. She could throw a stone as far as a boy and climb a tree more quickly and neatly than any of them. Zizi was the very spirit of youth and happiness. It did everyone good to see her.

Then Zizi's father bought her a bicycle, greatly increasing the pleasure of seeing her for a certain section of the community —i.e., every male verging upon adolescence up to those a couple of weeks away from senility. Because while riding her new bicycle Zizi wore her old jeans and the village in which she lived was, as I have pointed out, hilly. Quite a lot of motorists used to wait with engines running in the little *place* for Zizi to finish her shopping, and then jostle one another dangerously for the privilege of being the first to follow her up the hill.

When Zizi was sixteen—and looking every inch of twenty-two—her indulgent father bought her a mobilette, a rather

heavier bicycle with a small engine attached to the front wheel,
This made things easier for the motorists in the little *place*, in
that they only had to drive up the hill in orderly procession to
get the benefit of Zizi coming down. For Zizi, out of respect for
her approaching womanhood, had substituted a mini-skirt for
the childish jeans, and would come dashing down the hill with
her feet on the front mudguard and the little engine doing all
the work. Despite her speed, however there was only one
mishap during this time. As Zizi swooped past him Jean-Pierre,
the rising young electrician, drove straight into a telegraph
pole, which fell on top of his modest Dauphine and more or
less wrote it off.

Perhaps as a tribute to his interest in her Zizi married him a
year later, to the disappointment of her admirers in the *place*,
in that all that they saw of her from now on was her admittedly
lovely face appearing just above the door of Jean-Pierre's
new car—a fiercely souped-up Renault painted French racing
blue and covered with rallying stripes. For the first time they
regretted the unique distribution of Zizi's components. A long
body, they felt, would have given them a clearer view, while
short legs no longer mattered.

Then, one day, into the little *place* there drove slowly but
somehow erratically an exceedingly battered little *Deux
Cheveaux*, that Citroen maid-of-all-work which is really no more
than a motorised pram. It drove straight into the wall of the
butcher's shop and came to a halt. Out of it stepped Zizi, a
thought self-consciously. She disappeared into the shop, came
out with her parcels, got into the *Deux Cheveaux*, backed into a
new Peugeot 404, came forward again, knocked over the delivery boy's motorbike, reversed into a Renault 4 Vitesses and
left the *place* as slowly as she had entered it.

In the next three days the priest lost his rear bumper,
Madame Machou's new wooden gates were levelled and the
fishman suffered a puncture in the refrigeration section of his

van. Surprisingly enough, however, Zizi's *Deaux Cheveaux* kept going—a splendid advertisement for the marque.

Most people were agreed that once again Zizi's components were at fault, that a greater length of body would have given her a clearer view of the road ahead. But, in any case, everything was changed. The motorists in the *place*, instead of trying to get next to Zizi, now tried to keep as far away from her as they could. All day—and all night, too—the quiet hills resounded with the blare of nervous horn-blowing, and the occasional crash of breaking glass. But Nemesis was on his way from Paris, in the form of a huge furniture pantechnicon. Zizi, going up the hill, met it coming down, and finding no room to pass on either side charged it gallantly head on.

Zizi, a little older and more solid, is now driving a pram. In the pram is a charming little baby girl with her mother's tawny hair and golden eyes. She is also called Zizi.

Many of us oldsters in the little *place* hope we shall survive sufficiently long to welcome her first mobilette, though all of us are prepared to leave it at that.

One Out—All Out

THE electric razor ceased to work so abruptly that I knew at once what the trouble was. Probably, it had been coming on for some time. The armature electrode had simply detached itself from the wattage conductor, or some damn thing like that, and now I had half a face shaved and the last safety razor blade had been used yesterday for cleaning paint off a window.

I was putting the electric razor back in is case, with the intention of doing something about it some other time, when Madame called down in a fair fury from the upstairs bathroom, 'What's wrong with the pump?'

I'd been put out myself, so I called back with some asperity, 'What do you think is wrong with the pump? Give me a basis for diagnosis.'

'I can't hear you.'

'What's wrong with it?'

Women.

'There's no water in the tap and I can't rinse the laundry. Please go and look at it no wait, you needn't bother. There's no electricity. It's the strike.'

I should have known there was nothing wrong with my armature electrode. Of course it was the strike. We'd been waiting for it ever since the near-revolution of last May. The Syndicalists—the sinister French word for trades unionists—had done it again. It was March 11th., and there would be no electricity in the whole of France from 9 a.m. till 5 p.m.—

leaving me with half a face shaved and her with the laundry unrinsed, and both lavatories out of action. *Quel malheur*!

At about ten we went down to the Village to get some food and found that both the greengrocer and the butcher were in darkness, and because they were in darkness they'd closed their doors. It was something we hadn't allowed for, so we went home again.

I went down to the study to write some letters. It's always rather dark down there, as it faces east, so I switched on the table lamp and nothing happened. Then I noticed that nothing had happened either to the electric fire, which I'd just plugged in, so I went upstairs again and found that she'd erected the ironing-board in the kitchen and was shaking the iron, trying to make it work.

I was just about to unplug it, to do some running repairs to the armature electrode, when simultaneously we remembered the facts.

We went and sat in the livingroom for a while. There was no point in going down the hill again to get the newspapers because even if there had been any they wouldn't have arrived, and in any case the bar that sells them, being in darkness, would also be out of business.

I said, 'For jollity, let's put on the—never mind it doesn't matter.'

'The record-player?'

'Do not speak about it.'

At midday the clouds came piling up again from the east, the way they've been doing for weeks. She said, 'I'm terribly cold.'

I said, 'We've got no wood left and I'm sure the wood-merchant is in darkness, and shut, so we're done.'

'We could have the elec—' she said, and stopped.

There was a pleasant diversion during the afternoon. Geoff arrived with his pump. Thanks to the torrential rain that has been falling on the Côte d'Azur there is as much water in the plastic cover over the pool as there is in the pool itself, and I've been meaning for weeks to get rid of it.

We lifted the pump out of the boot of Geoff's car—it weighed a ton—and carried it into the garden. While Geoff primed the suction end of the hose I unrolled the other to the furthest corner of the garden and put the spout into the gutter that runs down the road. There were hundreds of gallons to be got rid of and I didn't want to get our bit of land soaked.

Then Geoff plugged in the pump, and nothing happened.

The three of us looked at one another without speaking and then went inside and in the cold, semi-darkness, had a drink. The ice in the fridge, of course, was nearly melted, but it served.

At five o'clock that evening the main pump sprang to life, so did Geoff's, the iron burnt a hole in the ironing-board and the following morning I found that the electric fire and the table-lamp had been on in the study all night.

General de Gaulle apparently made a great speech about the whole thing, but I haven't been able to get down to buy a paper.

Incredibly, the car battery is flat.

Hurried Money Scurry

To:
M. Ernie.
c/o Premium Bonds.
Lytham St. Annes.
ANGLETERRE.
Cher Monsieur,
        Thank you for your super letter of recent date.

Of course, when it arrived it got a bit mangled in my over-eager attempts to open it before it got away or dissolved into dust, but after I'd smoothed it out and the type had stopped leaping about I found it made jolly welcome reading.

Not, naturally enough, the long-desired Big Bonanza, but still it was good to learn from you that any winnings accruing to me from your national lottery could be paid without let, impost, tax or other hindrance directly to me here in France.

This was good news indeed, in that I had previously believed that any such winnings would naturally be impounded by the Chancellor of the Exchequer—is it Sir Stafford Cripps again? —on the grounds that the winner should not have left the country, but should have remained to pay it all back again by way of the increase in the cost of petrol, tobacco and drink.

But then, cher monsieur, I began to reflect upon the financial repercussions of suddenly finding myself in possession of—say —£25,000 of unearned income in the form of devalued sterling, floating, as it were, in limbo between the United Kingdom and France.

I must be frank with you. I got into an absolute panic. If, you see, you put your finger on me next week and I won the Big Prize and it was transferred to me here I would receive not £25,000 but something like £19,000—a loss equivalent to having two-thirds of one's blood drained out of the system.

Should it, perhaps, be left in England? Even invested in becoming stocks and shares? But say, then, that Sir Stafford suddenly thought of an even newer and more dynamic method of putting Britain back on her feet, with the result that the stock market fell like a stone the following day, would I be able to get my winnings out again before they became, in toto, worth 23/6d.—which, of course, would be before Capital Gains Tax, which Sir Stafford would have adjusted to be payable even upon the sale of bankrupt stock?

I tell you no lie. I began to sweat, to tremble, to get a genuine attack of the feverish fiscal shakes.

Might it not be better, I gibbered to myself, to swallow the devaluation loss and to have the £19,000 (approx) transferred to France? By the look of the new purchase tax increases here we're going to need every sou we can lay our hands on if we're not to spend Christmas in bed, by candlelight, eating home-grown, uncooked grass.

But how stable is the franc? If you remember, a number of English newspapers devalued it by about 10% last Sunday morning, a welcome move which would have increased my £25,000 to approximately £27,000 but by the following day it became apparent that the General hadn't heard about it, because he just carried on as before.

But does he know what he's doing? In one department at least he certainly does. From now on, if you leave France for 24 hours, you can take 50 francs—or about four pun ten—with you.

Say, now, that I believe—and I don't know *what* to believe—that the General is talking through his kepi, should I not begin to shift my £19,000 into Germany? I mean, *before* it's devalued

by perhaps 102%. But at 50 francs a time this would involve
an immense amount of motoring to the German frontier, and
back again for further supplies, and would probably defeat its
purpose in the end. After as few as ten trips one would certainly
require a whole new set of clothing and luggage, to judge by
the efficiency with which the French Customs are reported to
be searching these necessities. The Italian frontier is, of course,
much nearer, but those volatile, music loving people are said to
be laughing their heads off at travellers who offer their 50
francs for so much as half a plate of pasta.

Cher Ernie, I don't know *what* to do. I'm almost inclined to
urge you to declare a truce, to disregard my Bonds for a week—
or even two. But then you might get out of the habit of putting
your finger on them and I'd *never* win. Not that I ever have.

Do you know Dr. Kiesinger, personally? Could you slip
anything to come to him? Would it be *safe*? Reply by return.

Yours, *distrait comme tout.*

Patrice.